THE STRANGE STORY OF

THE QUANTUM

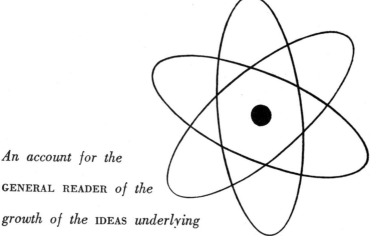

An account for the
GENERAL READER *of the*
growth of the IDEAS *underlying*

our present ATOMIC KNOWLEDGE

by BANESH HOFFMANN

DEPARTMENT OF MATHEMATICS, QUEENS COLLEGE, NEW YORK

Second Edition

DOVER PUBLICATIONS, INC., NEW YORK

Published in Canada by General Publishing Company, Ltd., 30 Lesmill Road, Don Mills, Toronto,
Ontario.
Published in the United Kingdom by Constable
and Company, Ltd., 10 Orange Street, London WC 2.

This Dover edition, first published in 1959, is an
unabridged and corrected republication of the work
originally published in 1947 by Harper and Brothers,
to which the author has added a new Postscript.

Standard Book Number: 486-20518-5
Library of Congress Catalog Card Number: 59-3821

Manufactured in the United States of America
Dover Publications, Inc.
180 Varick Street
New York, N.Y. 10014

My grateful thanks are due to my friends
Carl G. Hempel, Melber Phillips, and
Mark W. Zemansky, for many valuable sug-
gestions, and to the Institute for Advanced
Study where this book was begun.

B. HOFFMANN
The Institute for Advanced Study
Princeton, N. J., February, 1947

CONTENTS

vii

PREFACE

THIS book is designed to serve as a guide to those who would explore the theories by which the scientist seeks to comprehend the mysterious world of the atom. Nuclear fission and atomic bombs are not the whole of atomic science. Behind them lie extraordinary ideas and stirring events without which our understanding would be meager indeed.

The story of the quantum is the story of a confused and groping search for knowledge conducted by scientists of many lands on a front wider than the world of physics had ever seen before, illumined by flashes of insight, aided by accidents and guesses, and enlivened by coincidences such as one would expect to find only in fiction.

It is a story of turbulent revolution; of the undermining of a complacent physics that had long ruled a limited domain, of a subsequent interregnum predestined for destruction by its own inherent contradictions, and of the tempestuous emergence of a much chastened regime—Quantum Mechanics.

Though quantum mechanics rules newly discovered lands with a firm hand, its victory is not complete. What look like mere scratches on the brilliant surface of its domain reveal

themselves as fascinating crevasses betraying the darkness
within and luring the intrepid on to new adventure. Nor does
quantum mechanics hold undisputed sway but must share
dominion with that other rebel, relativity; and though, together, these two theories have led to the most penetrating
advances in our search for knowledge, they must yet remain
enemies. Their fundamental disagreement will not be resolved until both are subdued by a still more powerful theory
which will sweep away our present painfully won fancies concerning such things as space and time, and matter and radiation, and causality. The nature of this theory may only
be surmised, but that it will ultimately come is as certain as
that our civilization will endure—no more nor less.

What are those potent wraiths we call space and time,
without which our universe would be inconceivable? What
is that mystic essence, matter, which exists within us and
around in so many wondrous forms; which is at once the
servant and master of mind, and holds proud rank in the
hierarchy of the universe as a primary instrument of divine
creation? And what is that swiftest of celestial messengers,
radiation, which leaps the empty vastnesses of space with
lightning speed?

Though true answers there can be none, science is fated to
fret about such problems. It must forever spin tentative
theories around them, seeking to entrap therewith some
germ of truth upon which to poise its intricate superstructure.
The balance is delicate and every change sends tremors coursing through the edifice to its uttermost tip. The story of
relativity tells what happened to science when one provisional
theory of space and time yielded to another. The story of the
quantum tells of adventures which recently befell our theories

of matter and radiation, and of their unexpected consequences.

So abstract a matter as the quantum theory serves well as the basis for learned treatises whose pages overflow with the unfriendly symbols of higher mathematics. Here in this book is its story without mathematics yet without important omission of concept. Here too is a glimpse of the scientific theorist at work, pen and paper his implements, as he experiments with ideas. Not the least of his gifts is a talent for reaching valuable conclusions from what later prove to be faulty premises. For his insight is penetrating. Be it a hint here or a clue there, a crude analogy or a wild guess, he fashions working hypotheses from whatever material is at hand, and, with the divine gift of intuition for guide, courageously follows the faintest will-o'-the-wisp till it show him a way toward truth.

The magnificent rise of the quantum to a dominant position in modern science and philosophy is a story of drama and high adventure often well-nigh incredible. It is a chaotic tale, but amid the apparent chaos one gradually discerns a splendid architecture, each discovery, however seemingly irrelevant or nonsensical, falling cunningly into its appointed place till the whole intricate jigsaw is revealed as one of the major discoveries of the human mind.

THE STRANGE STORY OF THE QUANTUM

CHAPTER I

PROLOGUE

In the corner of a carefully darkened laboratory stands an electrical machine on which two small, shining spheres of metal frown menacingly at each other's proximity. It is a standard machine for making electric sparks to which one small addition has been made. Two metallic plates have been joined to the spheres by slender conducting rods, as if to add enormous ears to the two-eyed monster.

On another table stands a simple, almost closed hoop of stiff wire mounted on an insulating stand. For the experimenter the small gap in this hoop is the crucial part of the whole apparatus. If his surmise is correct, it is here that the secret will be revealed.

All is in readiness and the experimenter closes a switch to set the sparks crackling and spitting between the spheres. Turning his back on the flashing sparks he waits for his eyes to grow accustomed to the darkness. Is it imagination or does he really see a faint glow filling the gap in the ring? It is not easy to tell. It may be only a reflection. Gently he turns the screw that forces the ends together, and as the gap becomes narrower the glow seems to brighten. Closer yet, and closer, till the ends are almost touching. Now there can be no doubt. The experimenter breathes a sigh of relief. Tiny electric sparks are passing across the gap.

In so strangely simple a fashion did man first wittingly detect a radio signal.

This was in 1887, the experimenter a brilliant young German physicist, Heinrich Hertz.

The commercial value of this discovery was beyond estimation. Why, then, did so able a man as Hertz leave it for Marconi to capture the rich rewards of its exploitation?

It was not at all with the idea of inventing anything so practical as radio telegraphy that Hertz embarked on his epochal experiments, nor, perhaps, was radio telegraphy their most significant result. Hertz had set himself a task which had long baffled scientists: to test the truth of a highly mathematical theory concerning light, electricity, and magnetism proposed twenty-three years before by the Scottish physicist, James Clerk Maxwell. The thought of the commercial value of the work seems not to have troubled his mind at all, and this passion for pure research for its own sake was in a sense responsible for a most ironical situation. For without it Hertz might never have bothered about a seemingly trivial effect he had noticed in the course of his experiments. These experiments were everywhere hailed as brilliantly establishing the truth of Maxwell's theory on a rocklike basis of experimental fact. Yet the seemingly trivial phenomenon which he noticed was destined, in the hands of Einstein, to play a momentous part in the evolution of the quantum theory and thereby to aim at the theory of Maxwell a shattering blow from which it can never fully recover.

To appreciate the work of Maxwell and Hertz, and the

whole story of the quantum, we must first look briefly at some of the theories men have proposed about light.

Though there have been many notable Jewish scientists in modern times, the ancient Hebrew sages had no great instinct for scientific inquiry. Having disposed of the whole problem of light with the pronouncement *And God said, Let there be light: and there was light*, they quickly passed on to more important matters. Light, for them, was little more than the opposite of darkness, the circumstance of being able to see.

The Greeks, however, with surer scientific instinct, introduced a new idea of great importance. Realising there must be something bridging the distances between our eyes, the things we see, and the lamps illuminating them, they gave it objective reality and set about studying it and inventing theories about it. When the modern scientist talks about light he has in mind just this something. The distinction between the mere sensation of being able to see and the newer, more objective light is a significant one, being analogous to that between the sensation experienced when one is struck by a stone and the stone which actually traverses space to do the hitting.

Unhappily, the Greeks, after an auspicious start, became involved in conflicting theories. According to one of these, light was something that streamed out of the eyes like water from a hose, the idea being that we see an object by directing this stream of light to hit it; much as a blind man "sees" something by putting forth his hand to touch it. This theory would explain why we see only in the direction we are facing, and why we are unable to see with our eyes closed. But it cannot explain, for example, why we do not see in the dark. It was in an endeavor to meet such objections as this that the philosopher Plato produced a theory which, for sheer profusion of superfluous mechanism, is surely without equal. He demanded a triple interaction between three separate streams—one from the eyes, one from what is seen, and one from the lamp illuminating it! Plato's difficulty lay in his having got the direction wrong in the first place. According to modern ideas, when we see anything it is because light enters our eyes instead of leaving them, and the curious thing is that this view had already been vigorously put forward by the great Pythagoras more than a hundred years before Plato. The Pythagorean theory is simple. It holds that light is something that streams out from any luminous body in all directions, splashing against obstructions only to bounce off immediately. If, by chance, it ultimately enters our eyes, it produces in us the sensation of seeing the thing from which it last bounced.

Here we see nature at her most lavish and wasteful, making certain of catching our eyes by splashing her abundance of light in every possible direction, and showing none of the economy she delights to exhibit in the unerring precision of the grapefruit.

Of course, the problem of light is not at all solved by such a theory. Our troubles are just beginning. Every new discovery in science brings with it a host of new problems, just as the invention of the automobile brought with it gas stations, roads, garages, mechanics, and a thousand other subsidiary details. Here, for example, as soon as we realise that there must be something bridging the space between our eyes and what we see, a something to which we give the name "light," we open the floodgates for a torrent of questions about it; questions we could hardly have asked about it before we knew it was there to be asked about. For instance, what shape is it, and what size? Has it even shape or size? Is it material or ethereal? Has it weight? Does it jar anything it strikes? Is it hot or cold? How quickly does it move? Does it move at all? If it cannot penetrate thin cardboard, how does it manage to pass through glass? Are the different colors transmitted by the same light?—These and a multitude of even more embarrassing questions spring into being as soon as we discover that light exists.

As the tale of the quantum unfolds we shall come across answers to some of these questions, and shall enjoy the spectacle of Science again and again changing its mind. Other questions, whose answers lie outside the main stream of the story, will be heard of no more.

Two different theories arose to explain how light leaps across space to bear its message to our eyes. Let us begin by asking ourselves how we would move a stone that is out of reach. There are only two different methods, and they correspond to these two theories of light: one method is to throw something at the stone, and the other is to poke it with a stick.

The idea of throwing something was the inspiration for the first theory, the so-called particle, or corpuscular, theory. According to this theory, light consists of myriads of little specks, or "corpuscles," shot out by luminous bodies in all directions like the fragments from a continually bursting bomb.

The other theory, the wave, or undulatory, theory, was modeled on the stick method. But we must explain a little, since it is by no means obvious at first sight that prodding a stone with a stick can have anything at all to do with waves.

Let us remember though that, however rigid a stick may be, it is bound to be slightly compressible. And to bring this cardinal fact vividly before our minds let us pretend that the stick is made of fruit jello. Of course, such a stick would not move anything very heavy even if it did manage to hold up under its own weight, but in abstract science—and what we are talking about now is abstract science, however seemingly related to gastronomy—in abstract science it is the principle that is important, and the principle here is that the stick can never be perfectly rigid. So fruit jello it is. And to give it a fighting chance we must change the stone into a ping-pong ball.

What happens when we push on one end of the jello stick? The other end does not move immediately. Instead, a shudder begins to glide majestically down the stick, in the fullness of time reaches the other end, sets it in motion, and thus moves the ping-pong ball. This too, on a different scale, is what happens when we prod the stone with a steel rod. Now comes the crucial question. What was it that actually moved along the jello stick? A pulse, a mere shudder! Nothing so material as a thrown stone. Something as impalpable as the lingering grin of Alice's vanished Chesire cat. Yet something able, after all its travels, to move the ping-pong ball. If we reflect on this and distill its essence we arrive at the conclusion that light could carry its message from place to place by behaving like a wave. We come, in fact, to a wave theory of light.

But a wave in what? After all, a wave isn't just a wave. It must be a wave in something. Could it perhaps be a wave in the air? No, because light can travel through a vacuum. This fact alone shows that it cannot be a wave in any material medium; if there is anything material filling a vacuum it ceases to be a vacuum. Shall we therefore have to abandon the theory for want of a medium in which our waves can wave? By no means. No scientist is going to give up a promising theory for want of a simple hypothesis which no one can at the moment hope to disprove. All that is necessary is to say there must be some omnipresent, immaterial medium in which light is a wave, and to be careful to give it the dignity of an imposing name. It was called the luminiferous ether, and its sole reason for existing was to bolster the wave theory of light by lending it pictorial plausibility.

Here, then, we have two rival theories of light, the particle theory and the wave theory. Which one is correct?

The great Sir Isaac Newton, who made all his fundamental discoveries in dynamics, gravitation, the calculus, and many other phases of science in a mere dozen years of scientific activity, found time during that period to make significant advances in optics. Feeling that since waves spread around corners they could not explain why light travels in straight lines, he preferred to work with the particle theory. True, by this time many curious facts were known about light which did not seem to fit in with the particle picture. But Newton, a man of consummate genius, had little trouble in overcoming such difficulties. By the time he had finished he had succeeded, with only slight sacrifice of simplicity, in explaining practically everything then known about light. His particles were no longer characterless, however. Experimental facts had forced him to endow them with a curious ebb and flow in their power of being reflected. No longer was light analogous to the discharge of a blunderbuss, but rather to the pulsating flight of birds. We do not intend it as a pun when we say that this rhythmic pulsation will prove of interest in the light of later history.

Though the wave theory did not lack adherents in Newton's day, with so colossal a genius ranged against it it stood little chance of victory. The wave theorists, led by the Dutch physicist Huygens, based their chief hopes on the fact that particles ought to bounce off each other, while the experimental evidence pointed to the contrary fact that two beams of light could cross each other without suffering any damage. This alone, however, was scant basis for a theory to compete with Newton's pulsating particles.

After the death of Newton, new experimental discoveries were made about light and new techniques were invented to handle the difficult mathematics of wave motion. For all its ingenuity and simplicity, the particle theory fell upon evil days. The objection that waves would bend around corners was met when it was found that the waves of light were mere ripples measuring the fifty-thousandth part of an inch or so from crest to crest, for such minute ripples would not spread out noticeably. Of course, they would spread out a little, and it could be calculated that this would mean that light ought not to cast utterly sharp shadows but should produce definite patterns of fringes at the edges. Such fringes were actually known to exist even in Newton's time and Newton had been unable to account for them really satisfactorily. All the new evidence, both experimental and theoretical, led decisively away from the particle theory, and a hundred years or so after Newton's death the wave theory had been brought to so fine a degree of perfection by the Frenchman A. J. Fresnel as to reign supreme in place of its defeated rival. Fresnel developed the wave theory of light with such power and elegance that not any of the numerous intricate and beautiful experiments then known could escape elucidation by it. And if further proof were necessary that the particle theory was wrong it was found later in the decisive experiment of the Frenchman J. B. L. Foucault, in which the speed of light was actually measured in water. For it was on this point that the two theories differed decisively. In empty space light moves with the unthinkable speed of 186,000 miles per second. According to Newton, the speed in water should be even greater. The wave theory insisted it must be less. Science waited long for a Foucault to appear who could devise an experimental method for measuring such extreme speeds. When the experiment was made it showed that the speed in water was less than that in air by just the amount demanded by the wave theory. The particle theory's star had set, and from then on there was a new light in the heavens.

The evidence for the wave theory was already overwhelming. Yet it was to receive even more decisive support. Not long after the time of Fresnel there came a renaissance in the ancient and somewhat stagnant sciences of electricity and magnetism, a renaissance notable for the experimental researches of the Englishman M. Faraday, whose discovery of electromagnetic induction and invention of the dynamo laid the foundation for the present-day achievements of modern electrotechnology.

Faraday had little grasp of technical mathematics, a circumstance which to a lesser man in so mathematical a field would have proved an insurmountable obstacle. With Faraday, though, it was to be an asset, for it forced him to plow a lone furrow and invent a private pictorial system for explaining his experimental results to himself. This system, of an extreme simplicity, and peculiarly nonmathematical in appearance, was based on what Faraday called "tubes of force," and though at first somewhat ridiculed by the professional mathematicians of the time, it was to prove in some ways superior to their own systems. The mathematicians looked for the secret of electromagnetic effects mainly in the lumps of metal and the coils of wire that produced them. Faraday would have none of this. For him, in a real sense, nothing less than the whole universe was involved, the wires, magnets, and other material gadgets being rather insignificant incidents. The two points of view are nicely contrasted in the simple case of a magnet attracting a lump of iron. The mathematicians felt that the essential things here were the magnet, the iron, and the number of inches between them. For Faraday, on the other hand, the magnet was no ordinary lump of matter but a metal-bellied super-octopus stretching multitudinous, invisible tentacles in all directions to the uttermost ends of the world. It was by means of such tentacles, which Faraday called magnetic tubes of force, that the magnet was able to pull the iron to itself. The tentacles were the important thing for Faraday; they, and not the incidental bits of metal, were the ultimate reality.

With each experimental discovery Faraday brought new support for his ideas. Yet for long his tubes of force were felt to lack the precision needed for a mathematical theory. It was many years later that Maxwell became deeply interested in Faraday's ideas. From this interest was to spring one of the most beautiful generalizations in the whole history of physics, ranking, indeed, with Einstein's theory of relativity and with the quantum theory itself—the former firmly corroborating

its general form, and the latter corroding its very fundamentals!

Maxwell's first step was to translate the seemingly mystical
ideas of Faraday into the more familiar language of mathematics. This in itself was no small task, but when it was accomplished it revealed the idea of Faraday as of the very
quintessence of mathematical thought. From these labors
was born an important new physical concept, the field, which
was later to form the basis of Einstein's general theory of
relativity. The electromagnetic field is more or less the refined
mathematical form of Faraday's tubes of force. Instead of
thinking of space as filled with a multitude of separate tentacles we have to imagine that they have merged their identities within one smoothed-out and all-pervading essence of
tentacle, the electromagnetic field. The electromagnetic field
is to be thought of as an ultimate physical reality, the sum
of all those innumerable stresses and tensions whose effects
we may observe when a magnet attracts iron, when a dynamo
makes electric current, when an electric train moves, and
when a radio wave carries our voices around the world. The
ubiquitous seat of all these tensions was called the ether, but
to preserve a careful distinction between this new ether and
the luminiferous ether demanded by the wave theory of light
it was referred to as the electromagnetic ether.

Not content with translating Faraday's ideas into mathematical form, Maxwell went on to develop the mathematical
consequences of the theory and to extend its realm. Soon he
came to a contradiction. Evidently all was not well with the
theory, but what the remedy might be was not easy to determine. Various scientists sought for a cure, among them
Maxwell himself. So refined and mathematical had the theory

of electricity and magnetism become by now that when Maxwell arrived at a cure by sheer intuition based upon most
unreliable analogies, he produced a group of equations differing but slightly in external form from the old equations. But
not only did the new equations remove the contradiction,
they also carried a significant new implication. They required
that there should exist such things as electromagnetic waves,
that these waves should move with the speed of light, and
that they should have all the other major known physical
properties of light. In fact, they must be the very waves invented to explain all that was known about light. When it
was shown that the intricate details of Fresnel's brilliant
theories were contained without exception within the new
electromagnetic equations, the identification of electromagnetic waves with light waves seemed inevitable, and with it
the identification of the two ethers which scientists had been
at such pains to keep distinct.

Before the theory could be accepted it was necessary that
Maxwell's hypothetical electromagnetic waves should be produced electrically in the laboratory. This turned out to be
difficult, the difficulty being not so much in producing them
as in proving they had actually been produced. As the years
went by and no such waves were detected, physicists began
to have misgivings as to the validity of Maxwell's ideas, especially since they were based on rather loose analogies. No
matter how attractive Maxwell's theory might be on paper,
unless electromagnetic waves were actually detected in the
laboratory and their properties investigated it could at best
be regarded as no more than an extremely interesting though
rather dubious hypothesis.

Maxwell did not live to see the vindication of his theory.

It was not till seven years after his death that the electromagnetic waves he had predicted were first detected by Hertz.

The faint sparks crossing the gap in Hertz's simple hoop told only that electromagnetic disturbances were traversing the laboratory. To prove these disturbances were waves required careful investigation. Hertz probed their behavior by moving his hoop from place to place and observing how the intensity of the sparks varied. With sparks so dim this was no easy task, yet by such crude means did Hertz prove that the disturbances exhibited reflection and refraction and other wavelike characteristics, and measured their wavelength. Subsequent measurement showed they moved with the speed of light, thus removing any lingering doubts that they behaved exactly as Maxwell had foretold and were fundamentally identical with light waves. Not radio telegraphy but this was the true significance of Hertz's work, that it established the correctness of Maxwell's theory.

And this was no meager theory. We may well ask how it could so positively assert, in the face of the clear evidence of our senses, that radio waves and light waves are the same sort of thing. Their difference lies in the frequency of the waves, the rapidity with which they pulsate, the number of vibrations they make per second. Already in the older wave theory of light, and even in Newton's theory of pulsating particles, this had been the difference between the various colors. It was to be extended to other forms of radiation. When light waves are of low frequency they correspond to red light. As the rate of vibration increases, the color changes to orange, then yellow, and so on right through the colors of the rainbow up to violet.

But why stop at the ends of the visible spectrum? Let us

anticipate later events to give the complete picture. As we go higher and higher in frequency we come to the invisible light called ultraviolet, then to X rays, and finally to the gamma rays from radium and other radioactive substances, and to some of the constituents of the cosmic rays. As we go lower in frequency than the red light waves, we pass through the infrared rays, and the heat rays, and finally reach the radio waves of Maxwell and Hertz. All these different types of radiation were ultimately found to be the same thing, differing only in frequency of vibration: differing, so to speak, only as to color. And all were intimately linked in their properties with the phenomena of electricity and magnetism, and with the mechanics of Newton. It is this magnificent unification growing out of Maxwell's theory which gives some measure of its greatness.

With this superb conception to top the already considerable achievements of Newton's dynamics, science could well feel complacent. Had it not now reduced the workings of the universe to precise mathematical law? Had it not shown that the universe must pursue its appointed course through all eternity, the motions of its parts strictly determined according to immutable patterns of exquisite mathematical elegance? Had it not shown that each individual particle of matter, every tiny ripple of radiation, and every tremor of ethereal tension must fulfill to the last jot and tittle the sublime laws which man and his mathematics had at last made plain? Here indeed was reason to be proud. The mighty universe was controlled by known equations, its every motion theoretically predictable, its every action proceeding majestically by known laws from cause to effect. True, there were insurmountable practical difficulties to prevent the making of a complete prediction, but in theory, could the numberless observations and measurements be made and the staggering computations be performed, the whole inscrutable destiny of the universe would be revealed in every detail. Nothing essential remained to be discovered. The pioneering had been done and it was now only a matter of extending the details of what was already known. A few men with almost prophetic powers were able to discern the stealthy approach of distant storms, but their warnings did little to disturb the general equanimity. Physics was essentially solved, and was found to be a complete and elegant system. The physicist was content to cultivate his garden, unaware that he would soon be cast forth into the wilderness because Planck and his followers were about to taste of the bitter but life-giving tree of knowledge.

Long before Maxwell the particle theory of light had lost all reason for existence. With the wave theory of light turning up again independently from so unexpected a source as electromagnetism, the particle theory was surely dead.

Yet in 1887, in the very experiments that confirmed the existence of Maxwell's waves, Hertz had noticed a curious happening. It was so slight as to be hardly worthy of comment: merely that when light from the flashing sparks of his transmitting apparatus shone on the open ends of his hoop, the faint sparks in the gap came slightly more readily.

CHAPTER II

ACT I

THE QUANTUM IS CONCEIVED

In 1887 Hertz had noticed the curious fact that when ultra-violet light shone on his apparatus the sparks came slightly more readily.

Little could he realize that here within his grasp lay what still remains one of the clearest and most direct evidences we have for the existence of the quantum. The world was not yet ready to receive so precious a gift. The recognition of the quantum had to await the turn of the century, and when it came it was from a quite different quarter.

We now know how completely the quantum permeates all of existence. With the physicist it has become almost an obsession, haunting his every equation, dictating his every experiment, and leading him into long and not always fruit-ful argument with philosopher and priest on God and free will. Already its advent has revolutionized certain aspects of theoretical chemistry, and from chemistry it is but a short step to biology, the science of life itself. Yet with the ubiqui-tous quantum insistently giving the plainest possible hints of its existence, it was first hesitantly recognized in a field where its hints were somewhat obscure, and then partly as the result of a happy guess.

The quantum made its official bow to physics in connection with what is called the "violet catastrophe." This piquant title (which should have been reserved for one of the more lurid pulp magazine tales of a certain celebrated mathematician in one of our California colleges) was squandered by the physicists on a purely theoretical catastrophe—in both senses of the phrase.

The violet catastrophe consisted in this: if one calculated in what manner a body ought to glow when heated, one found a mathematical formula which implied that all energy should long ago have escaped from matter in a catastrophic burst of ultraviolet radiation.

The absence of any such occurrence was but one of the reasons for concluding that the formula was incorrect. Yet it was not wholly bad. Actually, for light of low frequency the results were good. It was for light of high frequency that the formula went on a rampage and preached mythical catastrophe.

Another line of attack on the problem of the glowing body led to a different mathematical formula, which successfully avoided the violet catastrophe, agreeing excellently with experiment for light of high frequency.

Did this solve the problem, then? By no means. For, while the first formula, excellent for low frequencies, was wrong for high, the second, which could not be better for high frequencies, proved unsatisfactory for the low. Two formulas, each only half right.

Such, in brief, was the state of affairs in this field when

Max Planck, professor of theoretical physics in Berlin, entered on his crucial series of researches.

Planck first indulged in a little pure guesswork. He tried the effect of various ways of maltreating the two imperfect formulas until in 1900 he hit upon a single mathematical formula which for low frequencies looked just like the first and for high frequencies just like the second. No really fundamental reasoning was involved here. It was largely empirical patchwork and opportunism, like making a single suit of clothes by borrowing the trousers from one person and the coat from another. By good fortune and excellent judgment Planck managed to get trousers and coat to match, the resulting suit being enormously more valuable than the coat and trousers separately.

The new formula—it is called the radiation formula—agreed splendidly with experiment. But now Planck found himself in the position of the schoolboy who, having managed with great cunning to steal a glance at the answers to the day's homework, is chagrined to find the problems nevertheless difficult. Planck was not altogether unprepared for his new task of finding some sort of theoretical justification for the formula he had so neatly contrived. Long and inconclusive investigations had convinced him that only something drastic could hope to save the situation. Armed with this vital conviction, he worked on the problem with such a fury of concentration that at the end of but a few weeks he had found the answer, an answer so unorthodox that only after the lapse of seventeen eventful years did it earn him the Nobel prize.

An accurate description of Planck's reasoning would carry us too far into mathematical abstractions, but something of the spirit of the work can perhaps be conveyed by a slightly simplified narrative which, though not a literal account of his argument, is at least artistically accurate in that some part of the general quality and flavor is preserved, as in a parable. Nor will great harm be done if the story is told as though it were in fact scrupulously exact.

There is a certain mathematical trick, invented by the Greeks, whereby a baffling smoothness is replaced by a series of minute jerks much more amenable to mathematical treatment. This trick, the foundation of the calculus, is a simple one in its general aspects. For example, if we try to calculate (not measure) the length of the circumference of a circle whose diameter is one inch we find the smooth circumference offers little mathematical foothold. We therefore creep up on the problem while it is not looking. We begin by calculating something that does afford a firm mental footing. Then, after reaching a secure position overlooking our slippery problem, we can suddenly jump on it without risk of being thrown.

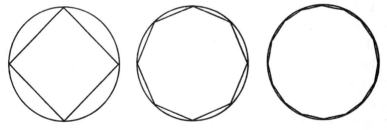

Thus, in the case of the circumference, we mark the circle into four, eight, sixteen, and so on equal parts and join the marks by straight lines, as shown. For each of these regular polygons it is possible to calculate the total perimeter, and it is obvious that as we take more and more smaller and smaller sides their total length will come closer and closer to the circumference of the circle. For instance, the total perimeter of the sixteen-sided figure is much closer to the circumference of the circle than is the sum of the sides of the square. What the mathematician does is to calculate the perimeter for a figure of some general number of sides. Then, after he has finished the general calculation, he suddenly smooths out the kinks by letting the number of sides in his formula increase without limit. In this way the unmanageable smoothness is not permitted to interfere with the details of the calculation until after the general formula is obtained. Incidentally, for a circle of one unit diameter the circumference is denoted by the Greek letter π, a number which keeps cropping up in the quantum theory. Approximately, π is 3.14, but if we try to write its exact value it objects violently and behaves like a supremely gifted trooper goaded beyond all human endurance, going on and on for ever without repeating itself:

$$\pi = 3.14159265358979323846264338327950 \ldots$$

Let us now return to Planck. Even before 1900 he had shown that for his particular purposes a lump of matter could be represented by innumerable particles rhythmically bobbing up and down. Some bobbed rapidly and others more slowly, all frequencies of oscillation being included. These oscillators, as Planck called them, had one simple job: to absorb heat and light energy by oscillating more violently and to give energy off again by letting the violence subside. They were just like children's swings, which on being pushed sweep through ever wider arcs; and they could hold energy as naturally as a sponge holds water.

A lump of matter absorbs energy by getting warm. Using his simple model, Planck calculated in what way matter would hold and give off heat and light at any temperature. Since he was dealing with smooth changes in the amount of energy absorbed or emitted, he employed the stratagem just described, replacing the smooth changes by jerky ones which he could calculate. On completing the calculations he found, as he had expected, that if he smoothed out the energy jerks in the conventional manner he was right back in the violet catastrophe. Now came the inestimable advantage of knowing the answer to the day's homework. From the start Planck had been prepared to seize any reasonable opportunity to indulge in some little impropriety if only it would give him the right answer, and here in his calculations he saw the chance he was seeking—a splendid chance, but a desperate one, for it invited an impropriety far from little. If he could bring himself to break with one of the most sacred traditions of physical theory by refusing to smooth out the energy jerks he could see a way to obtain the answer he knew agreed with experiment.

But such an idea was fantastic; it was like saying that a swing may swing with a sweep of one yard, or two yards, or three, or four, and so on, but not with a sweep of one and a quarter yards or any other in-between value. Even a child would realize how fantastic that sort of thing would be. Yet it did lead to the proper answer. . . .

If Planck let everything become smooth, the high frequencies would hog practically all the energy and cause catastrophe. Somehow he had to restrain them. Leaving the energy jerky did not in itself solve the problem, but it did afford a chance to exercise against the high frequencies a discrimination which was unconstitutional under the classical laws. For if Planck decreed that energy must be delivered in neat bundles, he could then go a step further and penalize the unruly higher frequencies by requiring them to gather far bigger bundles than the lower. A low frequency could then readily find the small quantity of energy needed for its bundle. But a high frequency would be much less likely to amass its onerous quota.

Using a convenient word, which had actually already appeared even in scientific literature in other connections, Planck called this bundle or quota a QUANTUM of energy.

To make the answer come out right, Planck found he must fix the quantum of energy for any particular frequency according to a definite rule—and, from the mathematical point of view, if hardly from the physical, a surprisingly simple one. Introducing a special quantity which he denoted by the letter h, he gave this illustrious and atomically explosive formula:

$$\text{quantum of energy} = h \text{ times frequency.}$$

The fundamental quantity h introduced by Planck and nowadays called Planck's constant is the proud ensign of the new physics and the central symbol of its defiance of the old order. From it tremendous events were to spring, yet it was hardly what one would call large. Its value was a mere

$$.000\ 000\ 000\ 000\ 000\ 000\ 000\ 000\ 000\ 006\ 6 \ldots$$

That h should be so extremely small meant that the energy jerks were incredibly feeble. Yet not to smooth them out entirely was something smacking of fire and brimstone and threatening peril to the immortal soul. This business of bundles of energy was unpardonable heresy, frightening to even the bravest physicist. Planck was by no means happy. And—an added terror to his situation—he knew he had had to contradict his own assumption of jerkiness in the course of his calculations. No wonder he strove desperately over the years to modify his theory, to see if he could possibly smooth out the jerks without sacrificing the answer.

But all was to no avail. The jerks do exist. The energy is absorbed in bundles. Energy quanta are a fundamental fact of nature. And to Max Planck had fallen the immortal honor of discovering them.

CHAPTER III

IT COMES TO LIGHT

For four years Planck's idea lived precariously, almost forsaken by its father. And then, in 1905, a certain clerk in the Swiss patent office at Berne made a bold and momentous pronouncement which was to revive Planck's languishing discovery and send it on its way strong and confident to its fateful assignation with Bohr in 1913.

Not long before, this same clerk had given a complete theoretical explanation of the so-called Brownian movement, and hardly four months after his brilliant resuscitation of Planck's discovery he announced a new theory concerning the electrodynamics of moving bodies, which we now know as the special theory of relativity. His name was Albert Einstein. So original and startling were his ideas that it was not till four years later that he was called away from his temporary haven in the patent office to join the university faculty in Zurich.

Einstein decided that Planck's idea must be made even more revolutionary than Planck himself had dared to imagine. According to Planck, energy could enter matter only in bundles; outside matter, where it took the form of radiation, it must obey the smooth laws set down by Maxwell. But Einstein showed that the two ideas would not balance each other, and showed further that the balance would be restored if radiation too consisted of bundles.

What was the net effect of these calculations? If anything, was it not rather damaging to Planck? Did it not imply that the upstart Planck conflicted with the well-established Maxwell? It required boldness and deep insight for young Einstein to say rather that it was Maxwell who conflicted with Planck.

Where Planck had demanded merely that matter should absorb or give off energy in bundles, Einstein now insisted that, even after escaping from matter, each quantum of energy, instead of behaving solely like a wave to please Maxwell, must somehow behave like a particle: a particle of light; what we call a photon.

It was a revolutionary proposal. But Einstein had some trump cards, prominent among them being the peculiar effect Hertz had noticed some twenty years before.

Much had been learned about this effect since then. In England J. J. Thomson had discovered the electron, and in Germany Lenard, who studied under Hertz, had tracked down the mechanism of the Hertz effect by showing that ultraviolet light is able to evaporate electrons from metal surfaces much as the sun's rays evaporate water from the ocean. It was this evaporation, now called the photoelectric effect, which caused the sparks to come more freely in Hertz's hoop.

Einstein gave a theory of the photoelectric effect which was an outstanding triumph for his new idea of light quanta. Unlike his theory of relativity, his theory of the photoelectric effect is easy to understand, as will be seen when we tell later how neatly it explained the many anomalies that had been observed in that effect. With the photoelectric effect lying at the basis of such things as the photoelectric cell, sound movies, and television, it is remarkable how many different by-products have come from Hertz's academic investigation of Maxwell's slight modification of the electromagnetic equations.

From Planck's jerkiness Einstein had developed the startling idea of a definite atomicity of energy. Imagine a sponge in a bathtub. We may liken it to a lump of glowing matter, and the water in the bath to the ether. According to Maxwell, when the sponge is squeezed it sends out its water in the usual way and causes waves in the bathtub. Planck's sponge is of a rarer sort. Indeed it is more like a bunch of grapes than a sponge, consisting of myriads of tiny balloons of various sizes, each full of water. When this sponge is squeezed, the balloons burst one after the other, each shooting out its contents in a single quick explosion—a bundle of water—and setting up waves of the same sort as Maxwell's. Einstein, however, took the sponge right out of the bathtub. He had no use for the water in the tub. When he squeezed his sponge gently, water fell from it like shimmering drops of rain. The jerkiness came not only from the inner mechanism of the sponge; it lay also in the very nature of the water itself, for the water stayed in the form of drops even after it left the sponge.

Einstein's was a very strange notion. To all intents and purposes it meant going back to the old particle theory of Newton. Even Newton's pulsations were there, playing an essential part. For it was the rate of these pulsations which, in the particle theory, was the frequency of the light, and frequency here had to play a double role. Not only must it distinguish the color of the photon, but according to Planck's rule, it must also determine its energy.

But who could believe so fantastic a theory? Had not the particle theory gone completely out of fashion, and with excellent reason, a hundred years before, and had not the wave theory thrust itself forward through two independent lines of research? How could a particle theory possibly hope to duplicate the indisputable triumphs of the wave theory? And who was this patent clerk anyhow? He was not a professor at a university. To go back to anything like the particle theory would be tantamount to admitting that the whole aesthetically satisfying and elaborately confirmed theory of electromagnetic phenomena was fundamentally false. Yet Einstein, not lightly and vaguely, but specifically and quantitatively, after deep thought and powerful argument, was actually proposing such a step.

Was it really so drastic, though? True, the wave theory had turned up independently in two different places, but Einstein was now merely evening the score to two all. And although for over a century all experiments had gone dead against the particle theory, had not such things as the violet catastrophe at last shown that Maxwell's theory also was headed for trouble? The fight was really not so unequal after all, even at the start.

The battle had first been engaged by Planck. Soon Einstein was making things more and more uncomfortable for the wave theory. While throwing off such items as the theory of relativity, he found time to return again and again to the attack, showing himself to be a mighty warrior, and exciting an increasing following of researchers. Repeatedly he and his followers brought forward new and important developments in support of the new view of light; no mere theoretical hair-splitting, but direct and simple explanations of experimental facts which had conveniently been avoided by the wave theory. But surpassing them all remained Einstein's explanation of the photoelectric effect.

At first sight there is something weird and almost miraculous about the photoelectric effect. Yet even from the point of view of Maxwell's theory it is natural that light should have some power over electrons, for Maxwell showed that light is electromagnetic, and an electromagnetic wave must surely influence so essentially electrical a particle as the electron. There was thus nothing really startling about the mere existence of the photoelectric effect. It was not that which baffled the wave theory. The surprise came when precise measurements were made of the speeds with which the electrons came off from the metal. If Maxwell's theory could be trusted, when the intensity, or amount, of light was increased the speeds of the electrons should be increased too. But what the experimenters found was something different. The speeds remained just the same as before. It was the number of electrons that increased. To increase the speeds the experimenters found they must increase not the intensity of the light but its frequency.

Here was a discrepancy between theory and experiment as serious as the violet catastrophe itself, even if less spectacular. Maxwell's theory was unable to explain the facts. Let us see how easily Einstein explained the whole thing with his photons.

Einstein looked upon the photoelectric experiment as a

sort of shooting gallery, with the photons as bullets and the teeming electrons in the metal as the ping-pong balls which bob so tantalizingly on fountains of water. To increase the intensity of the ultraviolet light is merely to increase the number of photons being shot out per second. This must inevitably result in more electrons being knocked out of the metal per second; which is precisely what the experimenters observed.

The effect of a change in frequency is explained just as elegantly. For, by Planck's rule, increasing the frequency of the light means increasing the energy of each photon, which is analogous to using heavier bullets. The higher the frequency, therefore, the bigger the jolt on the electron, and the bigger the jolt the greater the electron's speed. Again, this was exactly what the experimenters had observed.

When Einstein gave his explanation of the photoelectric effect, no really accurate measurements had been made of the way the speeds of the electrons changed with the frequency of the light. In 1906 he made a definite prediction on this point, a prediction based on his theory of photons and involving mathematics of such simplicity that any high school student can follow it. Later experiments, culminating in the classic researches of R. A. Millikan in America in 1915, established Einstein's formula with such precision and completeness that for a comparable verification of a scientific theory one must look to Hertz's confirmation of Maxwell's wave theory of light! It is curious that Einstein, who destroyed Newton's theory of gravitation with his general theory of relativity, should have played so large a part in resuscitating Newton's theory of light with his theory of photons.

Maxwell's theory was thoroughly outclassed over the photo-electric effect. And it came off just as badly against all the rest of Einstein's quantum ideas. Once the concept of the photon had been conceded, it was extraordinary how many well-known but little stressed phenomena, incomprehensible from the Maxwellian standpoint, were found to be in strict accordance with the new idea. From such diverse regions as photoluminescence, specific heats, and even photochemistry, Einstein and his followers gathered in the ammunition for their sallies. And with every advance the photon proved to be the simple key to just those problems which were unsolved by the theory of waves. It was not primarily for his monumental theory of relativity that Einstein at last received the Nobel prize in 1921 but for his services to theoretical physics in general, and specifically for his theory of the photoelectric effect. Two years later the Nobel prize was awarded to Millikan, whose precise measurements had so excellently confirmed Einstein's ideas.

Do not imagine that Einstein was the sworn foe of the theory of Maxwell. Far from it. Not only is the theory of relativity the apotheosis of Maxwell's concept of the field, but it also furnishes as beautiful a vindication of Maxwell's theory as Maxwell's theory itself gave to the wave theory of Huygens and Fresnel. The theory of relativity requires that every physical law shall fulfill a stringent condition. When the known laws of physics were tested against this condition they failed one after the other. The old ideas of measurement and simultaneity, of space and time, and of mass and energy, all had to go. The whole science of dynamics, including Newton's famous law of gravitation, had to be remodeled. Out of all that had been theoretical physics there were but two main survivors of the storm that was relativity. One of these was the set of conservation laws of mass, energy, and momentum, which said that none of these could be created or destroyed; but they were sorely changed from what they had been! The other was the Maxwell equations; they came through the storm unscathed, their form unaltered and proud —a supreme monument to the genius of Maxwell.

The conservation laws, which had been three distinct laws in the older physics, were fused together by relativity in indissoluble unity, nevermore to be separated. Mass was revealed as a form of energy; in fact, as the most concentratedly potent form of energy known, though its powers were latent and, at the time, there seemed little prospect of their being released. These powers were truly enormous. According to Einstein's formula, the energy inherent in a lump of matter would be calculated by multiplying its mass by the speed of light and the result again by the speed of light—a truly staggering quantity. Such was the measure of atomic energy. Part of this energy has now been tapped, with devastating results for the Japanese and momentous consequences for mankind. Yet even the atomic bomb for all its fury develops but a fraction of the full energy contained within its mass.

The ether was a special victim of Einstein's fell activities. Whichever way Einstein argued, whether for photons or for Maxwell and waves, the ether came off badly, losing, in fact, all reason for existing. In a pure particle theory of light, of course, an ether would be superfluous. But in the theory of relativity, which so smoothly incorporated Maxwell's electromagnetic waves, although these waves existed within the framework of the new theory they no longer needed an ether to wave in. That ubiquitous essence was superseded by space and time themselves, which had now taken on the power to bend and to transmit waves.

It was for the best that the ether, having served its purpose, should thus ultimately disappear from physics. In its heyday it had been a considerable nuisance, pretending to so many mutually contradictory characteristics that some of the finest scientific brains of the nineteenth century, brains that could ill be spared from the ramparts of scientific progress, were kept busy trying to devise ever more complicated mechanical models which would have properties bearing some faint resemblance to those of the ether as then conceived. The magnitude of this task will be appreciated from just one instance of the many discordant characteristics being claimed for the ether (not that the modern quantum theory does not succeed in encompassing phenomena which seem just as discordant!) Since it transmits light waves with prodigious speed, and these are of the special type called transverse waves, the ether cannot be a mere flabby jelly but must be a solid of extreme rigidity, far transcending the rigidity of the finest steel. Yet, although it must fill every nook and cranny of the universe, this stupendously rigid essence must offer not the slightest detectable resistance to the motions of the planets around the sun.

There is an element of tragedy in the life story of the ether. Its freely given services as midwife and nurse to the wave theory of light and to the concept of the field were of incalculable value to science. But after its charges had grown to man's estate it was ruthlessly, even joyfully, cast aside, its faith betrayed and its last days embittered by ridicule and ignominy. Now that it is gone it still remains unsung. Let us here give it decent burial, and on its tombstone let us inscribe a few appropriate lines:

First we had the luminiferous ether.
Then we had the electromagnetic ether.
And now we haven't e(i)ther.

CHAPTER IV

TWEEDLEDUM AND TWEEDLEDEE

WAVE or particle?

In the seventeenth century the particle theory of light had gained the upper hand, only to be deposed by the wave theory a hundred years later. And although, in the nineteenth century, the wave contracted with the electromagnetic theory of Maxwell a marriage so brilliant and strategic that the particle felt it must forever renounce hope of regaining its lost glory, the dawn of the twentieth century saw the beginnings of counterrevolution.

By now, though, the wave was well entrenched, and the resurrected particle, instead of bringing about a quick and decisive victory, succeeded only in plunging physics into civil war which was to drag on for more than a quarter of a century and to spread so rapidly that, by the time the armistice of 1927 was reached, the whole of physical science was irrevocably involved.

We have already watched the ominous gathering of the dark clouds of war, and the early skirmishes and flurries which herald the approaching storm. Now, the better to follow the restless, shifting tide of battle as the technical reports come

34

in, we must pause to inspect the main armaments of the rival theories, for they will be used later in strange places.

The armed might of the wave was great. It could well afford to keep the whole of the electromagnetic theory and the measurement of the speed of light in water as a second line of defense, for its more primitive armaments alone were seemingly overwhelming. We will look at but one of them.

The earliest insurrection against the particle theory of Newton had been armed with the fact that waves, but not particles, are able to pass through one another without injury, a phenomenon curiously named "interference."

The interference of waves was used to explain how it was that scientists could make two beams of light produce not more light but darkness. Imagine that we are shining two lamps on a bare white wall. The wall will be more or less evenly illuminated, and there will be nothing out of the ordinary to be noticed. Even if we could find lamps as small as pin points, as bright as a flash of lightning, and shining with light of a single frequency, there would still be nothing strange or unexpected to notice.

But now suppose that instead of using two different lamps we make one lamp do double duty—for instance, by letting it shine through two pin holes in a screen. Then the appearance of the wall would be different. No longer would it be uniformly illuminated. Instead, it would look something like the back of a miniature zebra, dark bands running across it in definite, regular patterns. These patterns are called interference patterns. The light has interfered to produce darkness. Interference patterns were discovered only after Newton's death. It would have been interesting to know what Newton and his particles could have done about them. They have yet to be explained in terms of any simple particle theory. For the wave theory, though, they were a conclusive vindication.

Imagine that an eccentric and cold-blooded millionaire in search of amusement insists on using you as his guinea pig in an experiment on the emotions. While his tough-looking bodyguards watch you closely, ready to pounce on you at the slightest sign of resistance or rebellion, the millionaire thrusts into your hands a bundle of thousand-dollar bills, only to snatch it away from your trembling fingers the very next moment; no sooner are you resigned to having lost it than he puts it in your hands again, but removes it before you can grasp it. If he continues his little game over and over again, there will be a considerable rhythmic fluctuation in your capital value which you might be justified in finding a trifle upsetting and not at all the best antidote for hypertension. Now comes the question, would two such millionaires be worse than one? Not necessarily. If they kept exactly in step with each other they would indeed be worse, for your capital assets would fluctuate twice as violently as before. But suppose they kept exactly out of step. Then at the precise moment when one millionaire proffered the money the other would snatch it up, the net result being that your capital assets would remain steady at their usual zero, or ten cents, or whatever they may have been. There would be no violent fluctuations, and you would find the presence of the two millionaires far more restful than that of either one alone. For when they are exactly out of step they interfere with each other, to produce no resulting effect.

In just the same way, if two waves of light always reach a certain spot exactly in step, their vibrations reinforce each other and produce a greater brightness than either would alone. But if they always arrive completely out of step, their vibrations oppose each other so that the net result is zero disturbance, or darkness.

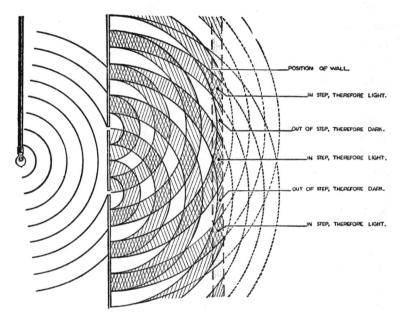

The diagram illustrates how a single lamp shining through two pinholes in a screen can produce interference patterns on a wall. Since the waves from the lamp reach the two pinholes simultaneously, the new waves that issue from the other side of the screen are in step with each other. At some places on the wall these waves are always in step and so produce brightness. At other places, though, the distances from the pinholes are just such as to make the waves always out of step. At those places their effects cancel and the result is darkness.

This is the wave theory's official explanation of interference patterns. It constitutes one of the most powerful and aggressive weapons of its armory in its struggle with the particle theory. For the particle theory to explain these interference effects one would have to imagine that the same particle went through both pinholes at once, surely a fantastic thing to imagine. Let us agree, then, with the physicists who made the experiments that when we find interference we are dealing with waves.

It was mainly by demonstrating interference that Hertz established the validity of Maxwell's theory of electromagnetic waves. It was interference too that showed that X rays are waves, for when X rays are passed through crystals they produce on a photographic plate a characteristic pattern which can be interpreted as the result of interference of waves disturbed by the regularly arranged atoms in the crystal. These X-ray diffraction patterns, as they are called, will appear again in our story.

Long before the coming of Maxwell, the formidable power of the wave theory had overwhelmed the hapless particle. It was a hundred and fifty years before the particle recovered sufficiently to begin the struggle anew. After so long a subjugation it could not hope to attack the wave openly on its own ground. It had to resort to cunning and seek out those obscure parts of the land where the power of the wave was practically nonexistent: shabby, uncultivated wildernesses which the wave had found too forbidding to develop. Defense was easy here, for the heavy artillery of the wave was powerless in such rough country, and the terrain was unsuited to the wave's refined and civilized habits. In such neglected places did the particle find refuge, and there it doggedly built for itself a new existence, tapping rich veins of gold beneath the barren ground. In the beginning the power of the particle was to that of the wave as a pebble is to the ocean, but so vigorous and rapid was its revival that soon its might stood forth like huge continents in the seven seas of physics, and in defense of its realm it produced many new weapons to match the colossal power of the wave's interference. By far the most dramatic of these was the photoelectric effect. Millikan's exact experimental verification of Einstein's formula could well be held in reserve by the particle to counter the second line of defense of the wave. There was something more primitive and striking about the photoelectric effect which it could well use in its front line: something just as conclusive for the particle as were the interference patterns for the wave, and something far more vivid.

Imagine that you have lined up along the seashore, near the water's edge, a long row of similar bottles. You leave them unattended while you have lunch. When you return you find one or two of them here and there lying on their sides, but most of them standing just where you had left them before lunch. Would you suppose that a huge wave had carefully taken pot shots at one or two of your bottles while just as carefully avoiding all the others? Such things occur only in the realm of the Disney cartoon. Much more likely that someone had passed by who could not resist the temptation to throw pebbles.

Now, what do we find in the photoelectric effect? The ultraviolet light does not knock electrons out of the metal from all over the surface at once. It knocks them out from here and there with no regularity or uniformity at all, except on the average. Could a wave cause such haphazard damage? There is no possibility here of interference patterns, for everything is uniform on the average. Surely only carelessly aimed particles could produce such sporadic and random effects. Surely light must be made of particles. If there is still doubt, we can call upon the evidence of very weak light. Suppose light was a wave. Then we could make the light so weak in intensity that, say, half an hour would elapse before enough energy had fallen on the whole surface to knock out a single electron. Since the light waves fall impartially over all of the surface, there would be no concentration upon any single electron. Weeks and weeks could pass before anything happened, and then suddenly, when enough energy had accumulated, electrons would start popping out right and left like an artillery barrage at zero hour. But nothing like this happens in practice. The electrons maintain a sporadic, desultory bombardment. And often this bombardment starts even before there has been time for the waves to produce any effect even if they were all concentrated on a single spot. With particles this is just what one would expect, for light is conceived of as a bombardment of photons. When the light is weak the bombardment is intermittent and the electronic response correspondingly sporadic.

Though perhaps lacking the thoroughness and solidity of organization of the wave, the photon has here a front-line weapon the equal of the wave's. Let us, for good measure, jump ahead of our story to include yet another evidence that light consists of particles. In 1911, after a dozen years of research, the English physicist C. T. R. Wilson invented that invaluable device the cloud chamber which renders visible the paths of individual electrons and other charged particles passing through it. In 1923 the American physicist A. H. Compton made a fundamental experiment which could be interpreted only on the theory that light bounces off electrons like one billiard ball off another. This was decidedly against the wave theory, of course, and two years later, working with Simon, Compton was able to observe in a cloud chamber the effects of individual impacts in this cosmic game of billiards. The electron tracks were directly observable, and the paths of the photons could be easily inferred from the positions of pairs of successively struck electrons. These various experiments left no doubt that individual photons were bouncing off electrons in strict accordance with the mathematical laws of impact. It is difficult to conceive of a game of billiards in which the cue ball is a wave, nor could one fulfill the laws of impact under such circumstances. Certainly light must consist of particles, then. And for this clear-cut demonstration of the fact Compton was to share the Nobel prize with Wilson in 1928.

We have now seen the primary armaments on the two sides. For the wave it was interference. For the particle it was the photoelectric effect and the manner in which light bounces off electrons. We have met other, more intricate weapons on either side, but for our purposes we may concern ourselves mainly with these, for they are basic and primitive. Let us now see what was the course of battle.

At the start all seemed confused, with first one side and then the other scoring the advantage. But the big guns of the wave theory proved to be lacking in mobility, and in its chosen terrain the particle was able to develop big guns of its own only to find them equally immobile. Soon the battle had degenerated into trench warfare, with neither side able successfully to attack the other. The photon could not capture the land of the wave and the wave could not overrun the domains of the photon. A stalemate set in with each side comfortably holding its own. The field of science was split between two warring camps, with the prospect neither of a quick decision nor of a reasonable compromise.

Science was not unfamiliar with situations where one theory covers one set of facts while a different theory explains another, but in previous cases there had been a plausible reason why this should happen. For instance, it did not cause anxiety that Maxwell's equations did not apply to gravitation, since nobody expected to find any link between electricity and gravitation at that particular level. But now physics was faced with an entirely new situation. The same entity, light, was at once a wave and a particle. How could one possibly imagine its proper size and shape? To produce interference it must be spread out, but to bounce off electrons it must be minutely localized. This was a fundamental dilemma, and the stalemate in the wave-photon battle meant that it must remain an engima to trouble the soul of every true physicist. It was intolerable that light should be two such contradictory things. It was against all the ideals and traditions of science to harbor such an unresolved dualism gnawing at its vital parts. Yet the evidence on either side could not be denied, and much water was to flow beneath the bridges before a way out of the quandary was to be found. The way out came as a result of a brilliant counterattack initiated by the wave theory, but to tell of this now would spoil the whole story. It is well that the reader should appreciate through personal experience the agony of the physicists of the period. They could but make the best of it, and went around with woebegone faces sadly complaining that on Mondays, Wednesdays, and Fridays they must look on light as a wave; on Tuesdays, Thursdays, and Saturdays, as a particle. On Sundays they simply prayed.

CHAPTER V

THE ATOM OF NIELS BOHR

IN 1911, with the battle between wave and particle locked in hopeless stalemate, a young man, Niels Bohr, crossed the gray seas from his native Denmark to continue his studies in England. After a year at Cambridge he proceeded to Manchester, where a man called Rutherford was professor of physics. Scientifically, Bohr was practically unknown. By professional standards his mathematical technique would not be called outstanding. But to him had been given the precious gifts of imagination and daring, and an instinct for physics having little need for intricate mathematics. Only to one such as he would the quantum reveal its next treasure, and today brilliant scientific virtuosos revere this quiet, unassuming man as the spiritual leader in atomic research. In 1922, the year after Einstein, he received the Nobel prize for physics. Two years before that, Bohr had become head of a newly created institute for theoretical physics in Copenhagen, which under the inspiration of his leadership was to grow into a world center of atomic research, attracting outstanding scientists of all nations, and exerting an incalculable influence on the headlong course of physical science.

This was Bohr. But who was the man Rutherford under whom he had come to study?

Back in the year 1895, a scant fifty years before the advent of the atomic bomb, when Rutherford was yet young, the German physicist W. K. Roentgen startled the world with his discovery of X rays. While experimenting with the passage of electrical discharges through gases, he came upon them more or less accidentally through noticing the glow they produced in fluorescent material lying near his apparatus. When the Nobel prizes were initiated in 1901, the award for physics went to Roentgen. It was not till 1912, however, that his X rays were shown to exhibit diffraction patterns characteristic of waves.

The discovery of X rays stimulated research in many directions, and a year later led the French scientist H. A. Becquerel to an accidental discovery of even greater moment.

Since the new-found rays caused fluorescence, it seemed possible they would be given off by substances which glow in the dark after exposure to light. Sure enough, Becquerel found that certain salts of uranium did give off these rays after being exposed to light. And then, by a fortunate chance, he discovered that they gave them off even without being exposed to light. This was indeed surprising. Uranium, then the heaviest known element, was found to be giving off penetrating rays spontaneously, generating them somehow without external aid. Becquerel had stumbled on what we now call radioactivity, and it proved to be profoundly disturbing. For, however minute in amount, here was energy without visible means of support. How did it arise? Whence did it come?

The radiations of Becquerel held irresistible fascination for the incomparable Marie Curie. Then a young and little-known scientist, she was later to become the only person ever to be awarded the Nobel prize twice, sharing the physics award with Becquerel and her husband Pierre Curie in 1903 and receiving the chemistry prize alone in 1911. Working in Paris, she and her husband were able, in 1898, to announce the existence of two new elements, each more powerfully radio-active than uranium. One they called polonium for Poland, the country of Marie's birth and the object of her fervid patriotism. The other they called radium. There followed four happy years of cruel, exhausting labor to distill by hand, from tons and tons of the refuse of uranium ore, a few small specks of precious radium salts.

The radioactivity of radium was almost incredible. It was by far the most active substance known, some two million times more potent than uranium. From within itself it sent forth an endless stream of energy in various forms: it glowed spontaneously in the dark and maintained itself a little warmer than its surroundings; it was later found to give off heavy radioactive gas previously unknown; it was to become a means of combating cancer; above all, it was to be recog-nized as a radiant witness to the terrible seething and boiling that goes on unceasingly within the very heart of matter. The amount of energy issuing from a speck of radium is extremely small, but such miniatures often foreshadow great events in science. Becquerel and the Curies had initiated the atomic age.

It is largely to Ernest Rutherford, and his English col-laborator F. Soddy, that we owe our understanding of the inner meaning of radioactivity. Working in Canada, they performed a masterly series of experiments which established the fundamental facts of the radioactive process and led them to formulate as early as 1903 a theory of the radioactive

disintegration and transmutation of atoms which, in its fundamentals, is accepted to this day. Rutherford was to make other
profound discoveries of enormous significance regarding the
atom which were to establish him as the greatest experimental physicist of the age. In 1908 he received the Nobel
prize for chemistry, the same award going to Soddy in 1921.

According to Rutherford and Soddy, radioactive atoms
were exploding and tearing themselves apart. Three types of
rays were being sent forth, dubbed α rays, β rays, and γ rays,
after the first three letters of the Greek alphabet. The γ rays
turned out to be X rays far more penetrating than those found
by Roentgen, and the β rays proved to be streams of electrons.
As for the α rays, they were fragments of radium which yet
were not fragments of radium. Though they resulted from the
explosion of the radium atoms, they were not radium but
atoms of a different substance, the inert and very light gas
helium, in what was called an ionized state because it carried
electric charge. When an atom of radium exploded, not
only were the α and β particles it gave off different substances,
but so too was the fragment that was left. This in turn was
radioactive and exploded, and its residues after it, in a long
chain of transmutations, ending in that most leaden of substances lead. A new vista had been opened up for science.

At that time little was known of the structure of the atom.
The discoverer of the electron, Thomson, who received the
Nobel prize in 1906, tentatively suggested that the atom
consisted of a ball of positive electricity with electrons em-
bedded in it like plums in a pudding. The swiftly flying
particles from radioactive substances were splendid weapons
with which to belabor the atom and wrest from it its secrets.
As Lenard, who received the Nobel prize in 1905, had al-

ready pointed out, the β particles passed so readily through atoms that there must be vast open spaces therein. But it was the α particles that provided the real puzzle, for they suffered violent collisions with atoms which could not possibly be accounted for on the basis of the Thomson model.

In 1911 Rutherford, now at Manchester and a Nobel laureate, proposed a new atomic model to account for these extraordinary collisions. He showed that the positive electric charge in the atom must be concentrated in a minute, heavy nucleus no more than the million-millionth of an inch across. It was impact with such compact, heavy nuclei that deflected the α particles so violently. The electrons in the atom, instead of being inside, as in the Thomson model, must be flying around the nucleus at relatively enormous distances, their combined negative electric charges just balancing the positive charge of the nucleus, and the whole structure bearing a marked resemblance to a miniature solar system.

Rutherford was no amateur or dilettante. He did not propose his model of the atom till he had mathematically proved the experimental evidence so compelling that escape from his conclusions seemed impossible. And indeed his atomic model is the basis of all our modern ideas of atomic structure. Yet the ability of the Rutherford atom to explain the results of experiment was offset by theoretical blemishes so deep rooted that only the most drastic treatment could hope to eradicate them. Let us tell here of two of these.

According to Maxwell's theory, a Rutherford atom would glow with light of all frequencies. Real atoms, on the contrary, had long been known to be very particular as to the frequencies of light they will permit themselves to be known by. Each element chooses for its own use, as a sort of trademark, a special group of frequencies of light, and no element ever successfully counterfeits the trademark of another. Take hydrogen, for example, the lightest and least complicated of all the elements. The spectroscopist, with his elaborate ritual for making hydrogen glow and examining its light with a prism, does not obtain a spectrum containing all the colors of the rainbow. Instead, he finds a complicated assortment of particular colors only. Since they appear as lines in the spectrum, they are called spectral lines. They can be arranged in families according to their position, and other indications. For hydrogen, on measuring the frequencies of the various lines and tabulating them in families, the spectroscopist finds the following array of numbers:

2,465,910,000,000,000	2,922,560,000,000,000	3,082,400,000,000,000	...
456,770,000,000,000	616,650,000,000,000	690,650,000,000,000	...
159,870,000,000,000	233,870,000,000,000	274,070,000,000,000	...
. . . .	. . . .	. . . .	

Now, these numbers, which can be measured with extraordinary precision, must certainly have a deep significance. They are the trademark of hydrogen, which no other element may usurp. It is inconceivable that the whole intricate system of individual trademarks should be a mere accident; that each trademark should be no more than a fortuitous aggregate of frequencies. These numbers must conceal the intimate personal secrets of hydrogen. But what is their message?

It was more than sixty years ago that an obscure Swiss schoolteacher, Johann Jakob Balmer, became fascinated by the riddle of these frequencies. In those days no more than four frequencies of the hydrogen atom were known, the others lying in the infrared and ultraviolet, beyond the visible part of the spectrum. From this meager material Balmer extracted an extraordinary formula which, though it accounted excellently for the four known frequencies, was altogether too strange to be readily accepted; its success might well have been merely accidental.

Balmer worked with wavelengths. Here is the kind of rule he discovered, modernized, and modified to refer to frequencies:

Take the mysterious number 3,287,870,000,000,000 and with it build a sort of irregular ladder leading down, the depths of its rungs being obtained by dividing this number respectively by 1, 4, 9, 16, 25, 36, . . . What makes Balmer's formula particularly intriguing is that these latter are not mysterious numbers. They are just the squares of the natural numbers 1, 2, 3, 4, 5, 6, . . . The ladder is shown on page 50.

What has all this to do with the hydrogen frequencies listed on page 48? Simply this: the frequencies in the first horizontal row are the distances from the first rung to the second, third, fourth, etc.; the frequencies in the second row are the distances from the second rung to the third, fourth, fifth, etc.

Balmer deduced the essentials of this remarkable rule from only the first four frequencies of the second row, and even went so far as to suggest that the other rows should exist too. The years passed. More and more frequencies were measured, not only for hydrogen but for the other elements as well. For all these, as the Swiss scientist W. Ritz showed in 1908, a ladder principle held. From the most meager data Balmer had performed the amazing feat of finding a key that was to fit the spectra of all the elements, and this he did so far before his time that he received no real recognition while he was alive.

The accuracy with which Balmer's concept fits the facts,

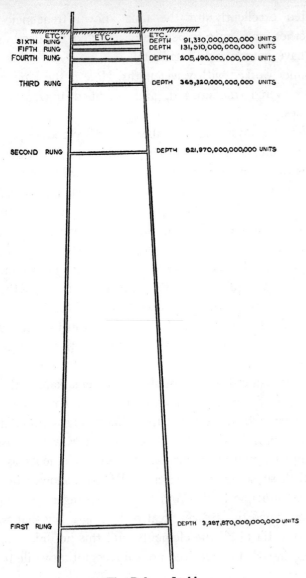

The Balmer Ladder

its profound simplicity and elegance, and the systematic entry of the natural numbers can leave no doubt that it represents a profoundly significant detail of the hidden anatomy of nature. It reveals a veritable backbone of the universe. Compared with it Rutherford's atom, with its insistence upon sending out light of all possible frequencies, appeared as a mere amoeba.

That was the first objection to Rutherford's atom. The second objection does not take so long to explain. It is simply that according to Maxwell's theory a Rutherford atom is an impossible structure. There is one essential difference between planets moving around a sun in the Newtonian manner and electrons flying around an atomic nucleus in conformity with the laws of electrodynamics. Planets move around the sun peacefully in elliptical orbits. But electrons moving around a nucleus would act very violently. They would not only send off their energy in waves of light of all frequencies, but while so doing would rapidly spiral right into the nucleus. If the atom were really something after the style of the Rutherford model, how could it possibly exist for any length of time?

With Rutherford's atom battling for its life against such catastrophic objections as these, young Bohr came forward in 1913 with proposals which can only be described as heroic. Had he not, he argued, come across this second objection before? Was it not something like the violet catastrophe all over again? And if so, did this not indicate that an injection of quantum would aid the ailing theory? And did not this in turn mean that the relatively slender experimental evidence for Rutherford's model must outweigh the voluminous, but already badly shaken evidence for the electromagnetic theory of Maxwell? True, Rutherford's own evidence was based directly on Maxwellian mathematics, but it is a poor research worker who insists on fully understanding his own intuition. And Bohr did not let himself be deterred by this small item; which was indeed good judgment, since the need to rely on Maxwell for want of something better still mocks the triumphs of some of the most modern theories of atomic physics.

To say that in setting up his quantum theory of the atom Bohr was guilty of outright plagiarism would be going too far. But there can surely be no reasonable doubt as to his tremendous unacknowledged obligation to that illustrious Englishman and instinctive diplomat, the Lord Chancellor in Iolanthe.

The incident is graphically related by W. S. Gilbert. The fairy Iolanthe, to save the Lord Chancellor from unsuspected bigamy and the plot of the opera from serious collapse, has revealed to him that she is his long-lost wife, thereby breaking a sacred vow of silence and incurring the penalty of death. While the remaining fairies look sorrowfully on, the fairy queen steels herself to pronounce dread sentence on the Iolanthe they all love.

But hold! What is this? A commotion! More; an interruption! As if impelled by a common, overmastering purpose utterly beyond their control, the ermine-clad peers of the realm are pouring in from all sides. What can it be that brings them here so powerfully? No one can tell! It is one of those unexplained mysteries so characteristic of opera. From their actions it would seem they are attracted by the sight of that veritable Prospero in incongruous formal attire down front whose baton so potently conjures forth music from the void. Certainly they have eyes for him alone. Whatever the reason, the advent of

the peers is more than opportune, for it gives the fairies courage
to try one last desperate chance to save their beloved Iolanthe:

THE FAIRY LEILA. Hold! If Iolanthe must die, so must we all;
    for, as she has sinned, so have we!
THE FAIRY QUEEN. What!
THE FAIRY CELIA. We are all fairy duchesses, marchionesses,
    countesses, viscountesses, and baronesses.
LORD MOUNTARARAT. It's our fault. They couldn't help
    themselves.
QUEEN. It seems they *have* helped themselves, and pretty
    freely, too! (*After a pause.*) You have all incurred death; but
    I can't slaughter the whole company! And yet (*unfolding a
    scroll*) the law is clear—every fairy must die who marries a
    mortal!
LORD CHANCELLOR. Allow me, as an old Equity draughts-
    man, to make a suggestion. The subtleties of the legal mind
    are equal to the emergency. The thing is really quite simple—
    the insertion of a single word will do it. Let it stand that
    every fairy shall die who *don't* marry a mortal, and there you
    are, out of your difficulty at once!
QUEEN. We like your humour. Very well!

Thus the Lord Chancellor. Thus also Niels Bohr. Faced with
a situation equally critical, he too inserted a single word.
Instead of the Maxwellian law

    Electrons moving around a nucleus radiate their energy and
spiral into the nucleus

he boldly suggested

    Electrons moving around a nucleus don't radiate their energy
and spiral into the nucleus.

Having taken this decisive step to save the Rutherford atom by
making the Maxwell theory the scapegoat for its impossibility,

he was now free to make any new rules he chose without feeling constrained by the ordinary niceties. With the way thus cleared, he replaced the Maxwellian laws by two rules expressly constructed to remove the other great blemish of the Rutherford atom—to supply it with a backbone.

Planets moving around the sun may move in circles of any size. The smaller the circle the greater the speed of the planet. To make a planet move in any particular circle with the sun as center, all that is necessary is to start it off properly with the right speed for that size of circle; nature will take care of everything else. According to the Newtonian theory of gravitation, all such circles are possible planetary orbits. Following the lead of Rutherford, Bohr took over this general picture; but he made some important amendments greatly restricting the freedom of the electrons moving around the nucleus. Only certain special orbits were to be permitted them, the rest being declared out of bounds. No longer could an electron roam fancy free wherever it wished but, more like a trolley car than a bus, it must keep strictly to the tracks laid down by Bohr, though, as we shall see shortly, it did have a little more freedom than the conventional trolley.

Bohr laid down his tracks with mathematical precision, using a formula that had actually been discovered the previous year by J. W. Nicholson of Oxford.

In retrospect we can see how direct an extension of Planck's fundamental idea was this new concept of Bohr's. Planck's great discovery amounted to a rule restricting the oscillations of his bobbing particles to certain permitted amplitudes only, all other amplitudes being forbidden. Bohr was simply applying the idea of forbidden motions to particles moving in circles instead of just oscillating up and down. So close was the parallel, in fact, that when, later on, the Nicholson-Bohr rule was being extended to more complex motions it was found that the rule for selecting permitted orbits and the rule governing the bobbing particles could be expressed by the same formula. This formula uses a rather pretty mathematical symbol, and it is so compact that little damage will be done if it is placed on exhibition:

$$\oint p \, dq = n \, h$$

For our purposes here it is to be examined no more closely than you or I would examine the usual museum piece, noting beauties and peculiarities, and listening to the smooth patter of the guide, but quickly passing on to the next exhibit. Here we notice Planck's constant h on the right, and also the letter n. This n, called a quantum number, is used to denote the natural numbers one after another in turn. The natural numbers are thus seen to be purposely inserted in the formula—a point to remember for later in our story.

For the professional mathematician this formula is packed with information, applicable to all sorts of circumstances. For the case of electrons moving in circles around a nucleus it can be interpreted as saying, in the jargon of Nicholson and Bohr, that the angular momentum of the electron must be an integral multiple of Planck's constant divided by twice π. But this is rather technical. Let us (with an eye to future developments) try to interpret more graphically the formula for the present case.

Imagine the electron orbits to be trolley tracks and yourself the contractor who must build them. Deciding to start with the tracks for an electron of some particular speed, you take from your pocket a well-thumbed copy of "Everybody's

Manual of Electron Trolley Track Construction" and look up the size of the circle needed for that speed. Next you telephone the factory and tell them to send you some track capable of withstanding the speed you are interested in. They send it to you in segments, and at once you realize that you have on your hands a harder problem than you thought. For, because of unusual manufacturing difficulties, the factory is able to make the track only in segments of a particular length depending on the speed to be withstood. The factory thus has a manual of its own, and it does not agree any too well with yours. For example, your track must have a length of seventeen units, but for this speed the factory can make track only in segments of three units length. Though flexible, the track is so tough as to be quite unbreakable; and three doesn't go into seventeen. What will you do? You cannot build the orbital track you want because there would be an overlap. You will simply have to declare that particular orbit impossible to construct. That is no way to make money, though. You will have to try your luck again with a different speed. This one turns out to need a circle of length twenty-five units, but the factory happens to construct track for it only in lengths of four units. Again the segments don't fit the circle, and you begin to despair of ever being able to construct an orbit. However, a systematic search of both manuals reveals several possibilities. There is one in which the segment happens to be exactly the length of the circumference. There is a larger one where the corresponding segment is just half the circumference. Another, yet larger, where the segments step around the circumference just three times. Another four times, and so on without end. These are the permitted orbits, all others being forbidden.

Though the above analogy may seem arbitrary, it is hardly

more so than was the original rule it describes. Restricting
the electrons to special orbits in this blunt way was Bohr's
first amendment. But nature would not let him stop there. The
restrictions on the freedom of the electrons were too drastic
and some outlet had to be provided for their natural wander-
lust. Bohr permitted an electron to jump the tracks whenever
it felt like it, so long as it did not play truant in the forbidden
zone but immediately lighted on another permitted orbit, there
dutifully to follow its rounds till once more seized with wander-
lust. The electron was now a cross between a trolley and a
flea.

This jumping permit was Bohr's second amendment. It
brings up a special question. Since each orbit belongs to a
different energy, an electron jumping from one orbit to another must either gain or lose energy in the process. What
happens, then, to the law of conservation of energy, which says
that energy may be neither created nor destroyed, but only
altered in form? Did Bohr renounce this law? Not at all. It
was the high spot of the theory. He could now bring in Einstein's idea of the photon. The energy an electron lost in a
jump was to be converted into a photon of light whose color,
or frequency, could be calculated by Planck's rule on page 22.
If the electron gained energy in the jump instead of losing it,
it did so by swallowing up a photon of the appropriate frequency instead of emitting one.

Such was Bohr's drastic theory of the atom. The first step
was to renounce Maxwell. The second was to forbid all orbits
but a select group. And the third was to allow jumps from one
orbit to another provided the energy differences were taken
care of by single photons. The theory was more a theory of the
electrons moving around the nucleus than of the atom as a

whole. And if we look at it carefully we perceive that it is really a direct transcription into theoretical terms of the formula discovered by Balmer and his followers. Actual atoms exist; therefore Bohr begins by renouncing Maxwell. They glow with only certain special frequencies which are differences between levels in a frequency ladder. Since frequencies become energies on being multiplied by h, Bohr boldly incorporates an energy ladder in his atom by allowing his electrons only certain special orbits, his orbits forming, as it were, an energy ladder with circular rungs.

Bohr's theory does give the hydrogen frequencies with astounding accuracy, but it was so direct a transcription of the Balmer formula that there could be little credit in such a performance as it stood. Were the matter to have rested there, scant attention might have been paid to so arbitrary and curiously unorthodox a theory. But there was a spectacular by-product which absolutely compelled attention. Let it be freely conceded that all those things about special frequencies and differences in levels had been so obviously and unashamedly shoved into the theory right from the start that it would have been surprising if they had not returned at the end. What made the theory famous overnight was the wonderful dividend it paid. For in explaining the hydrogen frequencies it incidentally gave a full explanation of the mysterious number 3,287,-870,000,000,000 from which we built our ladder. This was sheer profit. It was not something purposely injected into the theory at the start. It came out of the theory by a relatively roundabout route, and thereby established the theory as a paramount contribution to science. The number turns out to be a none too complicated arithmetical hodgepodge of simple physical constants and mathematical odds and ends. Here is the recipe for it which came out of Bohr's theory in 1913:

INGREDIENTS

Group I	Group II
(a) the mass of the electron	(e) Planck's constant, h, raised to the third power
(b) the charge of the electron, raised to the fourth power	
(c) the number 2	
(d) the number π, squared	

Instructions:

Multiply the ingredients of Group I together, and divide by Group II. The result will agree with experiment to within two one-hundredths of one per cent.

This is the sort of concoction physicists relish. Such close agreement with experiment would have been remarkable even for a well-established theory. Considering the radical character of Bohr's collection of new rules, it is little short of miraculous. Later developments will indicate, indeed, that it is rather miraculous!

Newton, who built such great things on the foundations laid by Galileo, was born the year Galileo died. Bohr was born in 1885, the year in which Balmer announced his formula.

CHAPTER VI

THE ATOM OF BOHR KNEELS

BOHR's bold thrust into the unknown was in the direct line of progress, conforming meticulously to the best revolutionary tradition. In defying Maxwell, Bohr did no more than follow the precedent of Planck and Einstein; in specifying the allowed orbits, no more than amplify Planck's original call to arms; in introducing the photons, no more than present Einstein's idea with further triumphs. His theory was the rallying point for scattered forces of revolution and his genius lay in instinctively knowing how to bring them together. Almost all the ingredients of the theory were the common property of hundreds of physicists. But there was only one Bohr.

The rise of Bohr's theory was meteoric. Almost immediately the energy ladder implied by its orbits was shown to have a direct physical existence by experiments of J. Franck and G. Hertz in Germany, which won them the Nobel prize in 1925. Success followed success with such rapidity, new theoretical results were so readily discovered and found to agree so closely with experiment, that in the general excitement the old controversy between wave and particle was almost forgotten. Bohr had opened up so attractive and fertile a region that few men could be drawn aside to contemplate the barren-looking wastes whereon the wave-particle war was still being fought.

All too soon, however, the world was plunged into another kind of war: a war of cannon and human blood, of primitive airplanes and heartbreak, of submarines and starvation, and mud-bespattered death. H. G. J. Moseley, finest of England's younger scientists, enlisted in the ranks and there was none with wisdom enough to say him nay. He was killed in the abortive attacks near Gallipoli in the Dardanelles, and it was Rutherford himself who wrote his obituary.

In the intensifying gloom the flame of abstract research still shone. War could not extinguish so elemental a fire. Archimedes, lost in abstract contemplation, had been oblivious of the enemy blow that killed him. With the Napoleonic Wars ravaging Europe, French engineers made the first accurate survey of the size of the earth. During World War I, Einstein perfected his theory of gravitation, the general theory of relativity. And while that war was still fresh in memory, an English expedition under A. S. Eddington journeyed to distant islands with delicate eclipse apparatus, tested Einstein's theory, confirmed its prediction of the bending of light rays, and announced to a war-weary world that England's former enemy was host to the greatest scientist of the age.

Despite war and its aftermath, Bohr's theory grew mightily in stature even as it increased in complexity, its circular orbits giving place to elliptic, and to other orbits of quite involved shapes. Of its many brilliant achievements let us briefly mention but a few.

In 1913, young Moseley, experimenting with X rays, found fundamental regularities implying that the nuclear charge increases by equal amounts from one element to another. This was a strong confirmation of Rutherford's atomic model, and the main details of Moseley's work were later shown to be an equally strong confirmation of Bohr's theory.

Though Planck's original formula was in excellent agreement with experiment, his oscillating particles were hardly more than a schematic representation of matter. In the old days nothing better had been available, but now the success of the Bohr atom emphasized that the theoretical basis of Planck's formula must be modernized to fit the new concepts of matter. The problem proved unexpectedly difficult, but in 1917, using general arguments, Einstein showed, among other notable things, not only that Planck's radiation formula could be derived in terms of atoms containing energy ladders, but also that the relationship Bohr had assumed between the energy jumps and the light was a necessary consequence, thus at one stroke confirming both Planck and Bohr.

Again, there is the Zeeman effect, for which the Dutch physicist P. Zeeman shared the Nobel prize in 1902 with his brilliant compatriot H. A. Lorentz. Back in 1896, influenced by the theories of Lorentz, Zeeman examined the light from glowing atoms situated in the field of a powerful magnet and found the spectral lines slightly broadened. Later, with more powerful equipment, he and others found that individual spectral lines split into groups of three, and even more. The Zeeman triplets could be accounted for on the Maxwellian theory of Lorentz, but not the more complex splitting. Though the Bohr theory also encountered difficulties with the more complex splitting, it was equal to the challenge of the triplets. In the normal hydrogen atom the allowed orbits were specified by the single quantum number n. It turned out that under the influence of a magnet the allowed orbits became more numerous so that two quantum numbers were required for each orbit. This led to a complete explanation of the Zeeman triplets—with one reservation.

The German physicist J. Stark, who received the Nobel prize in 1919, used electricity where Zeeman had used magnetism and found in 1913 that this led to an even greater complication of the spectral lines. What normally were single lines now became as many as thirty-two and more. In this instance the classical theory was powerless. It could give no explanation of the effect, a fact which made the Bohr theory's victory all the more impressive. For in 1916, in the midst of war, Schwartz-schild and Epstein independently showed that, with a third quantum number, Bohr's theory successfully accounted for the details of this intricate type of splitting. But here too we must make a reservation.

When normal atoms are studied spectroscopically with the most powerful instruments, the spectral lines are found to be individual bundles of fine lines forming so delicate a pattern that only the bundle as a whole could be discerned with the earlier instruments. No external influence such as a magnet is causing this so-called fine structure. Where can the Bohr theory find an internal influence that will explain it?

It was the German theorist A. Sommerfeld who, in the war year 1915, found a solution to the problem. The key was relativity. According to relativity, the faster anything moves the heavier it becomes. Applying this principle to the Bohr atom, Sommerfeld found a formula, agreeing excellently with experiment, which has since been bettered in minor details only, and incidentally has had a remarkable history. But to this paragraph too we must add a reservation.

The reservation in all these cases is the same. The original

Bohr theory was too prolific, giving not only the spectral lines that were observed but also many more that were not. This prodigality was a fundamental defect of the theory, showing up not just in these specific instances but in almost every case to which it was applied, for the theory was palpably incomplete. Though it could speak of the frequencies of spectral lines, it had nothing to say about such things as their relative brightness. Yet certainly some lines were bright and others dim. Bohr needed some way of calculating their brightness, and with luck this could solve his other problem by assigning zero brightness to the unwanted lines. By 1918 he had found a makeshift device having its origins in his original work of 1913. In typical fashion he simply added another rule to his theory, this one an embarrassing mixture of classical and quantum concepts which he called the correspondence principle. It will be described in a later chapter. Like most of Bohr's ideas, it accomplished its purpose surprisingly well. Among other things, it successfully aborted almost all the unwanted spectral lines, and it proved an indispensable guiding principle in later, more tentative explorations.

The long list of achievements of the Bohr theory is an imposing tribute to the greatness of its founder. With a framework of the most elementary sort, and using comparatively simple mathematical machinery, the theory went far beyond its immediate aims to yield results transcending all reasonable expectation. From the beginning it took over the leadership in the study of spectra, inspired and co-ordinated a multitude of new atomic experiments, and provided valuable clues for the analysis and interpretation of their results. Above all, it established the quantum in its rightful position at the forefront of fundamental progress in physics, and with every advance in knowledge the historical importance of Bohr's theory in the evolution of scientific thought becomes more and more apparent.

That a theory capable of such signal achievements should be destined to be swept aside a mere dozen years after its inception is but an indication of the stupendous pace of scientific progress in this particular era. The Bohr theory had made a dangerous and implacable enemy. It had not only dealt slightingly with the powerful wave-particle controversy, but had added insult to injury by attracting attention away from it. The wave-particle controversy could not forgive so serious an affront. Rightly considering itself the center of physics, it could never tolerate a theory which thus usurped its place, and its revenge, though long in maturing, was ultimately swift and devastating. Had the Bohr theory been able to destroy the controversy it might have survived to this day. But, beyond endorsing the photon without vanquishing the wave, it had sedulously cultivated its own garden and carefully avoided any constructive action toward ending the warfare between them. This isolationism was a fundamental weakness in its structure, which left it a defenseless prey to disharmony and inner contradiction.

With the earliest symptoms of the coming dissolution, the first disquieting failures of theoretical predictions to match experimental data, conveniently hidden behind the vagueness of the correspondence principle, the life of the theory was artificially extended beyond its natural span. But the seeds of dissolution lay within the theory itself, and the inevitable harvest could not long be delayed. Serious discrepancies began to appear between theory and experiment, which could no longer be masked by an appeal to the correspondence principle. Some indeed had no relation to that principle at all, for they affected the quantum numbers themselves, the very sinews of the theory. Spectroscopic observation showed that these quantum numbers should often not be whole numbers at all but whole numbers and a half, something the Bohr theory was unable to explain. Worse still, where the Bohr theory would indicate that a quantum number be squared, as 4×4, the spectroscopic analysists insisted it be multiplied by the next higher number, as 4×5. The experimenters found they could produce anomalous Zeeman effects in which the triplets became intricate clusters of lines defying all the arts of the Bohr theory. Even the normal spectra of atoms with more than one electron proved too much for the theory. Soon it became clear that the tide of success had definitely turned and was running strongly against Bohr, as once it had run against Maxwell. By 1924 the Bohr theory was reduced to living precariously from day to day, continually changing its position in a desperate effort to shield itself from the increasing blows of adversity, while all the time only vaguely aware of the identity of the enemy fundamentally responsible for its plight. And then suddenly it was gone.

Such was the Bohr theory of the atom. It was bravely constructed and bravely kept alive in a rapidly changing world. That it should so soon have been swept aside cannot dim its glory. In defeat, as in victory, the Bohr theory had greatness, for its very success had prompted the swift discovery of those discrepancies which were the ostensible reason for its downfall, and the newer theories which later took over its high position could never have survived the uncertain days of their infancy had not the Bohr theory already explored the wilderness and prepared the way for them.

Toward the end the Bohr theory staged a minor rally. We know the planets spin upon their axes as they travel round the sun. In 1925, S. Goudsmit and G. E. Uhlenbeck suggested that electrons do likewise as they go around a nucleus, for if this idea were hedged about with many artificial restrictions it would perform remarkable feats, helping to explain the anomalies of complex spectra and, surprisingly, even accounting for the fine structure of the hydrogen lines—without the use of relativity. This last was quite a puzzler. Was the fine structure due to relativity, as Sommerfeld had demonstrated a decade before, or was it due to the electron spin?

For all its makeshift character, the spin commanded respect, yet it came too late to have more than a negligible effect on the fortunes of the Bohr theory that had spawned it. Its effect on physics, however, was to be anything but negligible. It was later found that a sort of spin must be ascribed to practically every type of fundamental particle in the universe. One of its many services may be told here.

Scientists had already been forced to allow each electron three quantum numbers. The spin introduced a fourth. This was of real interest, for the Austrian theorist Wolfgang Pauli had long sought a fourth quantum number for reasons of his own. In his early twenties, Pauli wrote a technical account of the theory of relativity which contained more than Einstein himself knew about the details of the theory, on Einstein's own, enthusiastic admission. Later, Pauli did important work in quantum physics, and during his researches hit upon a curious fact which, though obviously of the deepest significance, could not be fitted into Bohr's theory except as yet another special rule. This rule, for which he received the Nobel prize in 1945, is simplicity itself to state. It says that no two electrons may have the same set of four quantum numbers. It is as if the Bohr atom were a large city where electrons live in separate apartments. Each apartment has a different address, one quantum number indicating the street, another the house, a third the floor, and the fourth the apartment. These four quantum numbers are, then, the complete address of each apartment, and Pauli's principle is a regulation against over-crowding. Indeed, it is technically referred to as the exclusion principle. Because of it only one electron at a time may inhabit an apartment, another electron being forbidden entry until the first moves out. When Pauli discovered this rule, the electron still had only three quantum numbers, and he had had to attach a fourth to it arbitrarily. The discovery of the spin showed how all four numbers could naturally belong to the electron. With the exclusion principle it was at last possible to explain the physical basis of the periodic table of the elements discovered by the Russian chemist Mendeleev and refined by Moseley. Hitherto, in physical theory, whenever anything was constrained in any way, force could always be assumed as an explanation. In the Pauli principle there could be no question of ordinary forces. Here were influences of a quite new type. It was as if the electrons were politely told they might not enter and meekly obeyed; somewhat as if, instead of using the police force to prevent overcrowding, one should hang out a sign saying MEASLES or MUMPS.

The Pauli principle is basic in all modern theoretical investigations. It applies to other particles than electrons, and is known to be linked with tremendous effects within the nucleus. Its validity is the essential reason why chemistry is what it is. In nature, for the particles to which it applies, no exception to it has ever been found; in science, no complete explanation.

Though discovered in the reign of Bohr, the spin of the electron and the Pauli principle belong to a later epoch. They were unable to stem the adverse tide, and the Bohr theory is now a memory. But Bohr will not pass from our tale. Like Einstein, he has another part to play in the strange story of the quantum.

And now the time has come to lower the curtain on Act I. As Marcellus would have put it, something was rotten in the state of the theory from Denmark.

INTERMEZZO

AUTHOR'S WARNING TO THE READER

So FAR, at least, our story has preserved some semblance of orderliness. We have seen the stately rise of classical physics, culminating in Hertz's complete vindication of Maxwell's theory; the beginning of the revolution instigated by Planck; its ominous spread under the leadership of Einstein; and the unprecedented stalemate to which it degenerated. Meanwhile we have followed the fortunes of the Bohr theory of the atom from its meteoric rise to its swift decline, dragging science down with it into chaotic uncertainty.

If, however, all this has seemed to be the opposite of progress, if it has seemed to be more a headlong succession of patchworks and contradictory theories built upon shifting quicksands than a serious and continued advance in our understanding of nature, if it has seemed to destroy forever all faith in the sagacity and rationality of scientists, and in all reliance on a scientific method leading to such gross contradictions, then indeed will the events to come seem at times utterly grotesque and fanciful. For now the pace suddenly quickens. Not the atom but the theory of the atom is about to explode.

What happens next is so fast and furious that for a time all continuity is lost and physics becomes a boiling maelstrom of outlandish ideas in which only the keenest minds can distinguish the gold from the dross. Professional physicists, swept off their feet by the swift currents, were carried they knew not where, and it was years before the survivors recovered sufficiently to see, with the beginnings of perspective, that what had so overwhelmed their science had been the convulsive birth pangs of a new and greater era.

If you have read thus far, there is no dignified way of escape left to you. You have paid your fare, and climbed to the highest peak of the roller-coaster. You have therefore let yourself in for the inevitable consequences. It is no use trying to back out. You had warning in the preface of what to expect, and if contemplation of the heights there described now makes you giddy and apprehensive, I cannot accept responsibility. The going will be rough, but I can promise you excitement aplenty. So hold tight to your seat and hope for the best. We are about to push off into vertiginous space.

CHAPTER VIII

ACT II

THE EXPLOITS OF THE

REVOLUTIONARY PRINCE

As Bohr well knew, not all the early successes of his theory could hide its insufficiencies, for he had poured the new wine of the quantum into bottles that were old. Because it was a heady wine, men did not resist it, but drinking deep strode forth to conquer realms where once they had feared to tread. Far they went on their path of conquest, reckless of their resources. And when at last the bottles broke, they found themselves deep in alien land, confused, leaderless, and without inspiration.

The confusion was vividly expressed by the German physicist Max Born .Toward the end of 1924 he completed a book on atomic theory. Though all he had to tell was contained within it, he called it Volume I. Now, why should he call it Volume I when there was nothing to be put into Volume II? Because he was so sure the Bohr theory was doomed and some entirely new system must arise to take its place he proposed to devote Volume II to this as yet unborn theory— provided he was still alive when it appeared!

It did appear, and he did write his Volume II—much

sooner than he expected. Not only was a large portion of the new theory to originate within the year right under his nose, and not only was he to be a significant contributor to its growth and interpretation, but even as he was yet engaged in writing his Volume I the first shot was fired in the wild rioting that heralded the new age of physics.

It was to Prince Louis de Broglie, member of an old noble French family, that the honor fell of ushering in the revolution. His work had its roots in ideas he had published as early as 1922, and his fundamental manuscript was submitted to the scientific press in December of 1923, almost a year before the appearance of Born's Volume I. But de Broglie's work was then unrecognized. Nor was it to form the basis of Born's Volume II. What went into Volume II must wait till a later chapter. For, as we have warned, the story is about to become complex.

While scientists were still struggling ahead under the leadership of the ailing Bohr theory, de Broglie chose to rummage quietly among the ideas of Einstein's theory of relativity. His primary interest was with light rather than matter, but in the course of his reflections he had had the idea of endowing the photon with intrinsic mass. Though the concept of a photon possessing such mass is not now accepted, it led de Broglie to a discovery of the first magnitude, for such a photon has kinship with a particle of matter and its mathematical development suggested important parallels.

In view of the accumulation of evidence, argued de Broglie, it would be stupid to pretend there are no photons in light. Nor can one deny that there is also a wave. The two must coexist. Moreover, in relativity light and matter are linked together, for both appear therein as forms of energy. Bearing these things in mind, we can make a little chain of relations which carries curious implications. According to relativity, mass is one of the embodiments of energy. According to Planck's rule, energy is h times frequency. Here, then, is our chain:

Particles of matter have mass.

Mass is energy.

Energy implies frequency.

Frequency implies pulsation.

So, catching our breath for a moment, we conclude that particles have pulsation.

Let us proceed:

Pulsating particles are suspiciously like photons.

Photons are related to light waves.

Therefore matter should be related to "matter waves."

Or, to put it briefly, what's sauce for the goose is sauce for the gander.

But it is risky to convict on such slender suspicion alone. Assuming that a rhythmically pulsating material corpuscle is accompanied by a wave simply because it pulsates would be like assuming that a rhythmically breathing Marine corporal was accompanied by a Wave when perhaps he was only thinking of one. De Broglie must have grounds more relative than this.

Relativity can play queer tricks, and de Broglie found in it many suggestive connections between particles and waves. To follow one of his lines of reasoning we must know one particular fact about the theory of relativity, which we shall mention when the time comes. Let us forget for the moment about waves and concentrate on pulsating particles. Surprisingly enough, we can determine their exact rate of pulsation. It comes right out of our chain of relations. We know the mass. Multiply by the square of the speed of light and, according to Einstein's law, it becomes the energy. Divide this energy by Planck's constant and lo! it becomes the frequency. For all its fanciful quality, our chain of relations was a precise one mathematically. From it we have created a picture of a particle with a definite rate of pulsation.

Concentrate now on the pure pulsation. If we write down the usual mathematical expression for such a pulsation we can interpret it in two ways: either as a bottled-up heartbeat or else as a spread-out pulsation. This gave de Broglie some assurance that there would be no mathematical contradiction if he used both interpretations at once. Thus he assumed that a particle at rest not only possessed a localized heartbeat but also was accompanied by a widespread pulsation forever in step with it and extending over all the universe. This pulsation was as if a whole ocean were rising and falling like the floor of some vast elevator; there were no waves in the ordinary sense, just a steady rise and fall. Is this fantastic? Undoubtedly! But no more so than the Planck-Einstein photon, or the Bohr atom, or a host of other things already met and to come. (Please do not look over the side of the roller-coaster now. It is so high up. You really must try to get used to the new sensations in physics.)

So, a particle at rest is now to be regarded as immersed in a widespread pulsation which is everywhere in step simultaneously.

Did we say simultaneously? Relativity will not like that. The first thing relativity ever did in its life was to attack the meaning of the word. It will prick up its ears. It will launch forth upon an impassioned platform speech, insisting that simultaneity is relative. "Just you start moving that pulsating particle of yours," it will say, "and then see how your simultaneity goes all haywire! Don't say I didn't tell you!"

We had better explain about poor old relativity. It really does feel this simultaneity business strongly. That was how it began—undermining the concept of simultaneity. If one person saw two things far apart happen at the same time, that did not mean, according to relativity, that another person would agree they happened at the same time too. In fact, if the two things were far apart, and if one of the men was moving relative to the other, then the two men would definitely have to disagree. Out of this fundamental discovery Einstein developed the whole of his theory of relativity, with all its paradoxical consequences, including the result that no signals of any sort could travel faster than light. In the old days, if a pulsation was everywhere in step simultaneously it was everywhere in step simultaneously, and no nonsense. But under relativity, as soon as either the particle or the scientist begins to move, the whole scheme of simultaneity becomes warped. In 1905 Einstein, like Hamlet so long before him, had cried to the world:

"The time is out of joint . . ."

and perturbed physicists, grumbling as they stirred from their comfort and complacency, had continued in garbled form:

" . . . O cursed spite,
That ever [he] was born to set it right!"

De Broglie knew the idiosyncrasies of relativity. Einstein had given precise mathematical formulas for the warping of simultaneity. De Broglie could now apply them to his pulsations and find out what happened when the particle moved.

And what happened was—they turned into waves.

Perhaps we can see in a general way how this came about. To do so we have to keep in mind that, according to relativity, motion warps simultaneity. Naturally, merely to say that motion warps simultaneity is not to explain the inner niceties of the theory of relativity, nor even to make the notion of warped simultaneity itself at all pleasant or easy to accept. Our purpose, though, is to try to see how the pulsations turn into waves. Let us therefore take the warped simultaneity on trust and see by what method it accomplishes the conversion.

Imagine a series of corks A, B, C, D, E floating at equal intervals on the heaving "elevator floor" ocean. Here is how they will appear at various times:

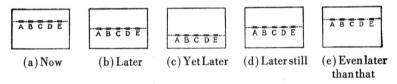

(a) Now   (b) Later   (c) Yet Later   (d) Later still   (e) Even later than that

They move up and down in step, of course, always keeping level with one another. But suppose we were so shortsighted that we had to get right up close to a cork before we could see it clearly. Then we could no longer take an over-all view of the situation. We would have to snatch a fleeting glance at A, rush on to look at B, hasten to C, then to D, and finally to E, all while the ocean was heaving up and down. What sort of impression would we have of the disposition of the corks? We would see A as in diagram (a), B as in (b), C as in (c), and so on. We would imagine the corks something like this:

We would, in fact, picture the ocean surface as having a wavy shape.

This waviness comes, of course, from our not observing the five corks simultaneously. But what about relativity? Does it not say that, if I am moving past the corks and you are not, my idea of simultaneous observation will not agree with yours? It says, in fact, that you will think my simultaneous, over-all observations were performed by some shortsighted messenger rushing from cork to cork with incredible speed. The warping of time will induce a warping of shape so that what you regard as the smooth surface of the ocean will appear to me to be ruffled; covered by waves. And it turns out that these waves will actually travel over the surface.

Travel is indeed the word, for they move much faster than light.

Did we say faster than light? Relativity will not like that. Now what? Oh. That is not so bad. Relativity objects only if actual energy is transmitted faster than light. Science had long known about waves moving faster than light which could not transmit energy that fast. They were called phase waves. They are quite all right.

But energy was nevertheless being transported, for the particle itself was now moving, and mass is energy. How did this link up with the phase waves? De Broglie discovered the connection.

If we take many trains of de Broglie waves, of slightly different speeds, they will add and subtract their effects in the manner of our old millionaire friends. Let us start them off in one direction, all being in step at one particular place. There will initially be an enormous wave at that place. But it will not remain there. De Broglie proved it would move along at a dignified pace, much slower than light. In fact, this towering, majestic composite wave would move—with the speed of the particle. That was the strange link between the slow-moving particle and the incredibly speedy wave.

De Broglie found other intimate connections between particles and their accompanying waves. For instance, the great French mathematician P. de Fermat had long ago reduced the laws of geometrical optics to the single all-embracing rule that a ray of light takes the path requiring the least time. Also, the laws of dynamics had been reduced to the single rule that any material system moves so as to use the least amount of a certain technical entity called action. On the one hand is a principle of least time; on the other, a principle of least action.

Now Planck's constant h happens to be a unit of this entity action. It is called, in fact, the quantum of action. De Broglie discovered that it acted as a bridge between wave and particle, the principle of least time for his matter waves being mathematically the same thing as the principle of least action for his particles.

How de Broglie's idea also gave a simple and striking picture of Bohr's rule for picking out the permitted orbits will be recounted in a later chapter.

Now, the professional physicist is a busy man. It is all he can do to keep abreast of the legitimate developments in his own special field. He is wary of cranks with worthless ideas designed to solve the universe. And there are many such. What was he, then, to make of de Broglie's suggestion? It was a pure speculation and quite fantastic for all its glib plausibility. It had no stunning triumph comparable to that which established Bohr's theory overnight: the recipe for the mysterious constant. Where was the experimental proof?

To be sure, de Broglie had been quite specific as to his waves, predicting that their wavelength must be equal to h divided by the mass and the speed of the particle. But was it likely that matter waves, if they existed, could have evaded the experimenter all these years? It was an interesting speculation, but surely nothing more. It was very pretty, and very subtle, amusing, and even striking; and elegant, and ingenious, and most astounding too. But was it physics? Where was the experimental proof?

If there was one man in all the world who might have anticipated de Broglie's discovery, that man was Einstein. For de Broglie's idea was the complement of his own idea of the photon and sprang from his own theory of relativity. Einstein had shown that light, long thought to be a wave, was like a particle. De Broglie had brought the argument round full circle by suggesting that matter, long thought to consist of particles, must be accompanied by waves and thus partake of their nature. Thus it was that when Einstein came across de Broglie's work he perceived at once its possible importance and placed behind it the weight of his far from negligible reputation. But still, where was the experimental proof?

In the Bell Telephone Laboratories, in New York City, C. J. Davisson had been conducting a series of experiments ever since 1921. What they had to do with telephones I do not know. But they did have to do with the bouncing of a stream of electrons off a lump of metal. In April of 1925 came an accident. Davisson, now aided by Germer, was bouncing electrons off a lump of nickel[1] in a high vacuum. While the

---

[1] I doubt that this is the missing link with telephones! [Postscript to a footnote, 1959: This jest is a casualty of time. I leave it in as a nostalgic reminder of better days when phone calls cost 5¢.]

lump of nickel was very hot, a flask of liquid air exploded in the laboratory, wrecked the apparatus, broke the vacuum, and let air rush in to ruin the carefully prepared surface of the nickel. The only practical method of cleaning the surface involved prolonged heating. Fortunately, Davisson and Germer, undaunted by the setback, repaired the damage, restored the surface of the nickel, and continued with their experiment.

Unknown to Davisson and Germer, the heat treatment had wrought a vast change in their nickel, fusing it into large crystals where before it had consisted of myriad small ones. Though the internal structure had thus been dramatically altered, there was no surface indication to betray the metamorphosis.

Davisson and Germer continued their interrupted experiment all unaware of the little game the gods of chance were so benevolently playing with them. With amazement they beheld the first of their new results. For here before their eyes were the typical patterns so long known to science as the diffraction patterns of X rays. But there had been no X rays—only electrons. The experiments had been started years before de Broglie announced his conclusions, and but for the accident of the exploded flask the experimenters would surely never have made their startling discovery. Now, Davisson was destined to receive the Nobel prize in 1937, and de Broglie before him in 1929. For these apparent X-ray diffraction patterns were the first direct experimental confirmation of de Broglie's theory. They showed that electrons behave like waves. And they showed more than this. They showed that electrons behave like the very waves de Broglie had predicted. For measurements proved that the wavelengths were just those which de Broglie had foretold. Thus was the confirmation placed upon a precise, quantitative basis. Here indeed was the inescapable experimental proof.

But what a curious situation for those irreconcilable feudists, the wave and the particle. And what a magnificent opening for the wave. Let us recall the primary armament of the wave, the armament on which it placed its main reliance, and which even the photon had had to concede was invincible: anything exhibiting interference must be a wave—the photon itself had admitted it, however reluctantly. All this time the photon had been glibly boasting it was a particle, just like any other particle—just like an electron, for instance. And now the electron, that ultimate particle par excellence, was found to be behaving like a wave.

One wonders what the particle might have thought had it been able to foresee one of the outcomes of all this. The ability of a microscope to disclose fine detail depends on the smallness of the wavelength of light used. Therefore the most powerful microscopes employed ultraviolet light. Since the wavelength of fast-moving electrons is thousands of times smaller than that of ultraviolet light, they gave promise of revealing far greater detail, a promise later amply fulfilled with the advent of the electron microscope.

After all the years of slow retreat the wave was able to launch the perfect counterattack: "So you say you're a particle, do you? Why, you don't even know what a particle is, you and your wonderful new ideas. What about your pal the electron? You said he was a particle. And look at him now. If you ask us, we think he's a wave. And we think you are too. In fact, we've been suspecting it all along, but you've been talking so almighty loud you almost began to get us confused." And, feeling much better after this outburst, the wave could sit back and enjoy its new-found happiness. But then the wave would begin to reflect—as light is prone to do. Was it really such a perfect counterattack? It had not won the war. It had not really attacked the photon in a vital part at all. It had merely extended the field of battle from the theory of light to the theory of matter as well. Of course, everyone had always supposed that matter belonged safely in the camp of the photon, and it was staggering to find it now in the forefront of the battle. But the citadel of the photon remained unvanquished. The wave could still not conquer the photoelectric effect. Nor could it conquer the cloud chamber tracks of the electron. Indeed, it had been a futile counterattack after all. In bringing the electron into the battle the wave had added the electron tracks to the armament of the particle. In claiming for itself the wavelike qualities of the electron it had driven the electron's particle-like qualities into the camp of the photon, there to set up combined headquarters for all particles. The civil war was now more desperate than ever, with almost all of fundamental physics inescapably involved.

Events were moving fast, however. Already diplomatic negotiations were under way in other quarters which, within a year, were to resolve the long-standing particle-wave controversy to the reasonable satisfaction of both parties.

But now we must go back in time to the days when de Broglie's theory was yet an unproved hypothesis struggling for recognition. De Broglie was living in a hectic age. By the time his ideas had been vindicated by Davisson and Germer they were already known to be physically not wholly tenable.

LAUNDRY LISTS ARE

DISCARDED

WITH de Broglie's idea awaiting recognition, Bohr's tottering theory still claimed the attention of physicists. Lacking a more definite guiding principle, men continued to use it in their calculations, calculations whose sole result seemed now to be to discredit it all the more.

Fortunate it was that de Broglie's idea paused awhile in the background. For otherwise two young researchers, the Dutchman H. A. Kramers and the German W. Heisenberg, might not have made a certain investigation based on the Bohr theory. This calculation came squarely up against the inadequacies of the Bohr correspondence principle, and gave to Heisenberg the germ of a noble and profound idea. Had the Bohr theory no more to its credit than this, that it revealed to Heisenberg the secret of its own weakness, and thus of the innermost weakness of all previous physics, it would still go down in history as a transcendental influence in the evolution of modern science.

In the days before Hitler there had gathered at the University of Göttingen, Germany, a brilliant and progressive group of men forming one of the chief glories of German science and mathematics, a group now scattered over the face of the earth. Here Max Born was a professor, and here Werner Heisenberg, a youngster in his early twenties, was a junior member of the faculty. Heisenberg's great discovery, when it came, was forbiddingly strange, far more so than the simultaneous pulsations of de Broglie. But Born, with keen discernment, could see in it the foundation of the new theory he had so confidently predicted in his Volume I. The initial idea came in 1925. Throwing all his resources into its development, and enlisting the aid of his colleague P. Jordan, Born was rewarded by being able to publish his Volume II, with Jordan, in 1930.

Like many another great idea, Heisenberg's is essentially simple. To show how it grew out of previous theories, however, we shall approach it circumspectly via the correspondence principle of Bohr, thus following its historic development. The correspondence principle will force us still further back into a branch of mathematics known as Fourier analysis. And this, in turn, will lead us to enter the unexpected realm of music.

Be it the limpid tone of a flute or the rich sonority of an orchestra, the fragile song of a distant nightingale or the awe-inspiring thunder of an atomic bomb, the unpretentious groove of a phonograph record will capture it and freeze it into a single wavy spiral. How does the simple groove perform such magic?

We can but say there is that about the nature of our ears and the wave character of sound which permits the most complicated noise thus to be recorded as a single wavy groove, and to be reproduced in all its finest detail through the trembling of a needle point which follows the groove's ripples.

Here is what the sound of an oboe looks like when captured by the groove of a phonograph record:

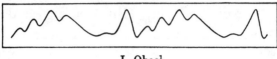

I. Oboe[1]

And here is the appearance of a clarinet:

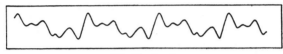

II. Clarinet[1]

When oboe and clarinet sound together the groove looks like this:

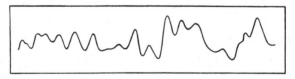

III. Oboe and Clarinet Together

Since sound is due to a wave motion, shape III is the net result of shapes I and II interfering with each other in the millionaire manner. Thus knowing I and II, it is a simple thing to create III. But it is a far different problem to discern in III the shapes of I and II which make it up. Though you may look as long as you will, you cannot unscramble the oboe from the clarinet.

But play the record on the phonograph, and your ear will know at once what instruments are being played, what notes they are playing, and what their relative loudness one to the

[1] After D. C. Miller, *The Science of Musical Sounds*, New York, The Macmillan Co.

other, and will even detect the extraneous noise of the needle scratching against the walls of the groove.

This is a veritable miracle of analysis! No sooner do we hear the record played than the whole complex analysis is completed. That intricate process which effectively tells us that III is I and II is completed by the ear instantaneously. And this is but a comparatively simple illustration. Think of the stupendous feats of analysis we perform every instant of our lives without so much as a thought. The complex jumble of air pulsations reaching our ears is automatically and effortlessly sorted out into constituents whose meanings are familiar. Amid the bustle and turmoil of traffic and the clamor of the crowd, we may yet discern the ticking of a watch and the sighing of the breeze. While engrossed in the majestic unfolding of a symphony, and delighting in the intricate interplay of instrument with instrument, we can still detect the rustle of our neighbor's program. These are incredible feats grown commonplace, dulled by repetition.

The mathematician, in his own way, though not with anything like this consummate ease, can perform comparable feats of analysis. He does not analyze a rhythmic pulsation into so much trumpet, and so much violin, and so much clarinet. His problems usually do not deal with music. He prefers, too, to use those simplest and purest types of oscillations, the sine waves, whose sound has the gentle sweetness of the flute and whose shape has the chaste rhythm of a ripple on still waters.

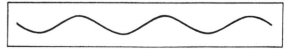

Sine Wave

Only a sine wave has a single frequency. All other rhythmic waves may be decomposed into constituent sine waves of different frequencies. This, which the French scientist J. Fourier had asserted over a hundred years ago, is the basis of what we call Fourier analysis. It can do things of which the ear is incapable. It can take the tone of a violin and find what are the pure sine tones that compose it. And when these pure tones are sounded together, as by striking a number of tuning forks, they do indeed combine to give the tone of the violin. The lowest frequency determines the pitch of the note. The higher frequencies, called harmonics, give the characteristic timbre. The frequencies of the harmonics are not haphazard, but are a whole number of times the lowest frequency. Thus, if the lowest frequency is one hundred per second, the frequencies of the harmonics will be two hundred per second, three hundred, four hundred, and so on, though not all these possible harmonics need be actually present in the tone.

Fourier analysis was the mathematical tool used by Bohr in his correspondence principle. The rhythm of the motion of a planet around the sun, when subjected to Fourier analysis, will yield a number of different pure frequencies. The same would be true for an electron traveling around the nucleus of an atom in one of the orbits permitted by Bohr. Now, we must be clear about one thing. These frequencies have nothing to do with the frequencies arising from the quantum jumps from orbit to orbit. They are quite different things. According to Bohr's theory, no one ever sees the frequencies in the individual orbits themselves. One sees only the frequencies corresponding to the energy jumps from one orbit to another. The two arise from different realms of physics in fact, for the former are classical and the latter quantum.

But Bohr felt he could make something out of this—

namely, his correspondence principle. Though the distances between permitted orbits grow larger and larger as we recede from the nucleus, the differences of their energies get smaller and smaller. Suppose an electron jumps from one orbit to another, both orbits being large. Measured as distance, the jump is tremendous. But measured in energy it is practically nil, and it is the energy jumps that are important, for it is they that produce the light we actually see. For large orbits, then, with the energy jumps almost smoothed out, should not quantum mechanics somehow merge with classical mechanics, and might there not therefore be some connection between the quantum jump frequencies and the classical orbital frequencies? You may justifiably ask why there should be, but it was Bohr's own theory and he could do what he liked with it. Besides, he had actually already noticed such a connection back in 1913 when he first set up his theory. In 1918, under the pressure of necessity, he pushed his connection beyond its legitimate range, applying it to large energy jumps and by his audacity managing to obtain the working rules for calculating intensities and the like for lack of which his theory was becoming seriously embarrassed. The trouble was that the classical frequencies no longer matched the quantum frequencies when the energy jumps were large—an obvious point which was nevertheless to prove of crucial significance, as we shall see. But Bohr managed in spite of this to set up a sort of correspondence between them through which he could take the classical results for such things as intensities and foist them on the allegedly corresponding quantum frequencies. This, in brief, was Bohr's celebrated correspondence principle. It was a most ingenious trick, and it really did work —to a surprising extent. But no one was at all deceived by it. It was a stopgap pure and simple. It was not really precise, even when expressed mathematically, and while some experiments would demand that it mean one thing, others would insist that it mean something else. Always its mixed parentage—part quantum, part classical—was a source of grave embarrassment. And its pathetic clinging to the irrelevant and discredited classical theory was indeed a confession of failure, or so it seemed at the time.

It is here that Heisenberg came in. The motion of a particle can be specified by two sets of quantities denoted by the letters p and q. These are no strangers to us. We have seen them before in the $\mathcal{f}$ rule for selecting Bohr's orbits. The q's denote the position of the particle, and the p's its momentum, that is, its velocity multiplied by its mass. For example, the wavelength of the de Broglie waves is h divided by p (cf. page 80).

According to Fourier, these p's and q's could be analyzed into their constituent pure sine waves, and according to the correspondence principle these should have relevance for the Bohr atom. But, in his work with Kramers, Heisenberg had found it necessary to tabulate the frequencies connected with the p's and q's, and this tabulation gave him a wonderful hint. For it was a square table.

This may not seem like a significant thing. But wait and see where it led Heisenberg and his followers.

Let us go back to the p's and the q's which describe the electron's motion. Having analyzed them into their constituent sine waves, we can make out a sort of laundry list of what is contained in them in the way of frequencies. Thus we shall say that in such and such a q there is first of all so much that is constant, then so much of this frequency, so much of that frequency, and so much of the other, beginning with the basic frequency and going right down the never-ending list of harmonics:

Laundry Mark: q		Laundry Mark: Smith	
Frequency	Amount	Article	Amount
Constant (i.e., zero per second)	4 units	Handkerchiefs	4
		Socks	8
1,000 per second	8 ”	Shirts	5
2,000 ” ”	5 ”	Towels	2
3,000 ” ”	2 ”	Dresses	0
4,000 ” ”	0 ”	Pajamas	1
5,000 ” ”	1 ”	Sheets	2
6,000 ” ”	2 ”	etc.	etc.
etc.	etc.		

This is an elegant arrangement (otherwise commercial laundries would not use it). With it we can see at once the exact constitution of each q and p, and mathematicians are always happy when they have arranged their data in some such way as this.

But Heisenberg was not satisfied. As he so clearly realized, this might be a good system for classical mechanics, but it was certainly wrong for the quantum, for the relations between the Fourier frequencies do not correspond to those between the frequencies of atomic spectra.

The starting point of Heisenberg's reconstruction of physics was the observation that the frequencies in the Balmer-Ritz ladders cannot be written down naturally in a sort of laundry list.

Every maker of road maps knows the reason why. Here is a small portion of a road map, done somewhat in the style of the map of the ocean in *The Hunting of the Snark*, though with somewhat more detail. It shows part of Route U.S. 1. Five towns are marked on the route, and the distances between them indicated by numbers. Though these five towns

NEW YORK	TRENTON	PHILADELPHIA		BALTIMORE	WASHINGTON
	61	32	105		41

Road Map: Route U.S. 1

lie on a single road, the natural way to make a mileage table
for them is like this:

MILEAGE TABLE

Name of Town	Baltimore	New York	Philadelphia	Trenton	Washington
Baltimore	0	198	105	137	41
New York	198	0	93	61	239
Philadelphia	105	93	0	32	146
Trenton	137	61	32	0	178
Washington	41	239	146	178	0

As a map, this portion of U.S. 1 is only a line, but as a
mileage table it is a square. Sometimes only the half of such
a table below the diagonal is shown, but that is an unessential
point. What is important is that it is not written out as a
single column in the manner of the laundry list.

Why does the mapmaker insist on a square or a triangle?
Is it that the shape appeals to his aesthetic sense? Is it, perhaps, that the square and triangle have deep occult significance in astrology and numerology? Of course, these might
possibly be secondary considerations, but the primary reason
is that the data to be tabulated cry out for such a form of
tabulation. It is the data that determine the tabulation to be
used. Who would ever think of making out a square laundry
list?

Or take, for instance, data on the number of gas stations
along the various stretches of U.S. 1. We could include this

in a square table just like the mileage table, even combining it with the mileage table, to give both mileage and gasoline data at once. More, we can actually force a square tabulation as against the formerly possible triangular tabulation by listing the number of gas stations on the right-hand side of the road only, something like this:

MILEAGE-GAS STATION TABLE

Name of Town	Baltimore	New York	etc.
Baltimore	250 gas stations in Baltimore	380 gas stations on the right in the 198 mile journey from Baltimore to New York	etc.
New York	420 gas stations on the right in the 198 mile journey from New York to Baltimore	525 gas stations in New York	etc.
etc.	etc.	etc.	etc.

From map data tabulator to physicist, in the present instance, is only a step. The physicist wishes to tabulate data concerning the frequencies contained in the Balmer-Ritz type of ladder. To make the analogy complete, let us therefore draw the frequency ladder for hydrogen as a road map:

| RUNG 1 | | RUNG 2 | RUNG 3 | RUNGS 4 5 6 etc. |

Road Map: The Balmer Ladder

The distances from rung to rung correspond to the distances from town to town, and the amplitudes of the various frequencies may be tabulated as were the numbers of gas stations above. While the motorist may wish to go from Trenton to Baltimore, the electron may wish to jump from the third rung to the fifth or from the tenth to the eighth. The main difference between the two cases is that there are now an infinite number of places on the map instead of only five. But this makes it all the more imperative to use a square tabulation.

Now, argued Heisenberg, all that we know for certain about atoms are such things as their respective trademarks, that is, the particular frequencies and intensities, etc., of the light which they give off. No one has ever seen the electron orbits. They are the purest fiction. We must forever put away such childish things, for they serve only to mislead us. Austerity must be the watchword for a theory come of age.

What, then, is left? With what shall we build our theory? With what construct a truer universe?

We must build it only out of the things we actually know: definite things such as the existence within atoms of Balmer-Ritz frequency ladders. A quantity like p, formerly mass times velocity, must now be made an endless square tabulation. So too must q, which tells a particle's position. All our atomic quantities, in fact, must be so represented.

This was a most stupendous undertaking. Look at what it must mean. For one thing, there was to be no facile insertion of particular frequencies right from the start. Though the rungs of the ladder might be labeled first rung, second rung, third rung, and so on, no definite frequencies were to be assigned to them beforehand. The theory itself must mathematically generate the correct frequencies and intensities for each particular situation. That was the stern requirement Heisenberg had in mind. At least it was specific. But what of the basic intangibility that bedeviled the whole Heisenbergian scheme? In the pre-Heisenberg era, though p's and q's could, if necessary, be analyzed into Fourier laundry lists, these lists could always be reconstituted into the p's and q's from which they had been obtained. But now there was no telling what the p's and q's might be. Only their square tabulations were to be known. These and these alone were the p's and q's, and not all the king's horses or all the king's men could put them back together again. With such grotesque p's and q's to represent the momenta of particles and their positions in space, what manner of universe was about to emerge? In it space and motion would certainly not be what they were. Yet Balmer and Ritz could not be denied. And Heisenberg was bold, for his age was twenty-three.

So Heisenberg renounced the Fourier laundry lists. For him a p or q could no longer be something familiar like this:

q	
Names of Ingredients	Amounts
The constant part	4 units of amplitude at zero frequency
The fundamental frequency	11 units of amplitude at 1000 per second
The first harmonic	6 units of amplitude at 2000 per second
The second harmonic	1 unit of amplitude at 3000 per second
The third harmonic	2 units of amplitude at 4000 per second
The fourth harmonic	0 units of amplitude at 5000 per second
etc.	etc.

Instead, it must be strange and square like this:

q				
Names of Rungs of Ladder	First Rung	Second Rung	Third Rung	etc.
First Rung	3 units of amplitude belonging to the first rung only	13 units of amplitude at the frequency belonging to a jump from rung 1 to rung 2	1 unit of amplitude at the frequency belonging to a jump from rung 1 to rung 3	etc.
Second Rung	8 units of amplitude at the frequency belonging to a jump from rung 2 to rung 1	0 units of amplitude belonging to the second rung only	0 units of amplitude at the frequency belonging to a jump from rung 2 to rung 3	etc.
Third Rung	5 units of amplitude at the frequency belonging to the jump from rung 3 to rung 1	2 units of amplitude at the frequency belonging to the jump from rung 3 to rung 2	4 units of amplitude at the frequency belonging to the third rung only	etc.
etc.	etc.	etc.	etc.	etc.

A different quantity, say p, would have exactly the same frequencies, but different amplitudes. Thus, if q corresponded to a table of gas stations, p would correspond to a similar table of some other data, say eating places, for the same route.

Heisenberg reconciled himself to carrying around his p's and q's thus broken up into little pieces, rattling in their square coffins like the bones of a skeleton—as, in a sense, they were to prove to be.

Anyone less valiant would have recoiled dismayed. Could such unwieldy filing cabinets, such unholy monstrosities, be the building bricks of the universe? Was it with these that one could reconstruct Nature—Nature, who had always shown herself ultimately simple?

Heisenberg had before him the unanswerable logic of experiment stripped to bare essentials. He must follow it wherever it should lead. If he now found himself cast off from the friendly shores of familiar mathematics, he was at least a voluntary exile, and if he would not turn back he had no other choice than to steer for the distant horizon. Ahead of him lay darkness, with never a star to guide his great adventure. Only from the fast-receding shore, where flickered the beacons kindled by men like Bohr, came a faint glimmer by which he might set his course. He had gone forth renouncing the warmth and comfort of the mainland, yet it was the mainland that served to light him on his way. For good or ill, he could not escape its influence. It was his native land, a part of his scientific heritage, never to be wholly forsaken or forgotten. The new world he sought must be fashioned in its image and lie in the direction it had pointed.

From the older p's and q's, penetrating equations of great power had been built to carry the exploration forward. There was much in these equations that was true. Could Heisenberg perhaps preserve their outward form, but build them out of his new, unwieldy tabulations?

In the old equations, the p's and q's were multiplied together. To re-create the form of these equations Heisenberg must discover how to multiply his square tabulations correspondingly. If he could but solve this crucial problem he would be able to set up the very equations of the older theories; equations already powerful, which would now be charged with a strange new element of unexplored potency.

How can one possibly "multiply" square tabulations together, though?

Well, in the older theories, when one multiplied p's and q's together, did one not by implication multiply together their corresponding laundry lists? Thus suppose one made a Fourier analysis of p, and another of q, and finally one of the quantity $p \times q$. Could not the laundry list of the last be regarded as the result of "multiplying" together the laundry lists of the original p and q? Not only is this not fantastic, it is a concept much used by mathematicians. The rule one obtains for multiplying laundry lists together is perhaps a little curious. But if it is curious rules we are going to worry about, what of the well-known rule we all use so glibly when merely adding simple fractions in arithmetic?

Just look, for instance, at what we must do to add $\frac{2}{13}$ and $\frac{3}{7}$. First we must multiply the 13 by the 7 to obtain a new denominator 91. So far it has been fairly straightforward, even though we may be surprised to find ourselves multiplying in the process of performing an addition. The next step is more complicated. To find the numerator we must go through the rather elaborate ritual of multiplying the 2 by the 7, and the 3 by the 13, and then adding the resulting 14 and 39 to obtain 53. Placing the numerator over the denominator, we finally have the answer, $\frac{53}{91}$. Surely if we can stomach so complicated a rule for merely adding fractions in arithmetic, we can hardly afford to be squeamish at a curious rule for multiplying laundry lists together in higher mathematics, or one to multiply Heisenberg's square tabulations in quantum physics.

Actually Heisenberg devised a rather natural rule, closely related to the rule for multiplying Fourier laundry lists. A simple illustration will suffice to show how the routine goes. Instead of the enormous tabulations of Heisenberg, let us use small ones having only four pigeonholes, and let us write in them only the amplitudes. Thus, let us pretend our p and q are as follows:

Rungs	1	2
1	2	4
2	6	8

$p \longrightarrow$

Rungs	1	2
1	1	3
2	5	7

$q \longrightarrow$

The result of the multiplication $p \times q$ must fit into the same sort of filing cabinet:

$p \times q \longrightarrow$

Rungs	1	2
1	?	?
2	?	?

What shall go into the various pigeonholes? Well, suppose, we multiplied the 6 in p by the 3 in q. That would give us an answer 18. But where should we put it? The 6 refers to a jump from rung 2 to rung 1, and the 3 to a jump from rung 1 to rung 2. So the 6×3, or 18, will refer to a jump from rung 2 to rung 1 and back again to rung 2. That is, it begins and ends in rung 2. It belongs, therefore, in the bottom right pigeonhole. What else will go in that pigeonhole? To find out, we look for all other possible double "jumps," the first in p and the second in q, whose net result is a "jump" from rung 2 to rung 2. There is the 8×7, since that is a "jump" from rung 2 to rung 2 followed by a similar "jump." But there are no others. So altogether in that pigeonhole we file away 18 and 56, that is, a total of 74.

What shall we put in the second pigeonhole on the top row? It must be something referring to a jump from rung 1 to rung 2. The 2×3 will belong there, since the 2 belongs to rung 1 only and the 3 belongs to a jump from rung 1 to rung 2. In this pigeonhole will also go the 4×7. So in all we have 6 plus 28, or 34, for that pigeonhole.

Proceeding in this way we readily find:

$$p \times q \longrightarrow$$

Rungs	1	2
1	22	34
2	46	74

Just for the fun of it, let us twist the p and q around and form the product $q \times p$. It is a routine, of course, since we have already worked out $p \times q$. But it is good practice. Here are our q and p, the same as before:

$$q \longrightarrow$$

Rungs	1	2
1	1	3
2	5	7

$$p \longrightarrow$$

Rungs	1	2
1	2	4
2	6	8

What will go in the top left pigeonhole? The 1×2 and the 3×6 of course, giving a total of 20.

But what is this? We got 22 before. Surely there is some mistake. Let us check it over. No, it is certainly 20. How about the previous 22? Perhaps that was in error. No, that too was correct; 2×1 and 4×5 certainly give 22. Let us try another pigeonhole, the one at the top right. Here we shall have the 1×4 and the 3×8, or 28 in all. Again it does not match the previous result. What a terrible situation! It can mean only one thing:

Heisenberg's rule of multiplication makes $p \times q$ different from $q \times p$.

Surely this result was a mockery of Heisenberg's highest hopes. A man less resolute, a man less deeply inspired, might have abandoned his quest on making this grotesque discovery.

By now the lone voyager was tiring and longing for news of the mainland. It had been difficult traveling alone the uncharted seas of tomorrow. He hastened to make an end of his present labors, and call on the aid of more experienced navigators. First, though, he must take soundings to test out the depth of his discovery. Hastily he made some preliminary calculations to see what manner of results might be obtained from it—and found an unmistakable portent of success. Scientists had long realized, from a variety of experimental results, that an oscillating particle such as Planck had envisaged could never be robbed of all its energy and thus be brought to rest. Half a quantum of energy must be forever imprisoned within it. Hitherto no theoretical explanation had been given for this residual energy. It was an arbitrary fact of nature that remained outside the basic theoretical structure of physics. Now Heisenberg, with his hasty calculations, found it was an automatic consequence of his new theory, and found too that the energy changes must occur in whole quanta just as before. This was his portent of success. This was the token evidence that his intuition had been correct.

In 1932 he was to receive the Nobel prize. It was now July of 1925.

Heisenberg returned to relate his scientific adventure. We may imagine him explaining to his eager hearers how strange his ideas had seemed even to himself; and when he came to tell of his rule for multiplying square tabulations and how $p \times q$ was not the same as $q \times p$ he may well have wondered how they would receive the news.

By one of those extraordinary scientific parallels which seem almost to be a part of some design, there had been a similar development more than half a century before. As Born pointed out to Heisenberg, in 1858 the English mathematician A. Cayley, investigating certain aspects of geometry, had invented a new and curious calculus, the calculus of matrices. These matrices were square tables of numbers obeying certain mathematical laws. When Heisenberg constructed his square tables and devised his special rules for handling them he was unwittingly rediscovering this matrix calculus. That such concepts should ultimately find their way into atomic physics in this particular manner was hitherto undreamed of, and revolutionary in the extreme.

This was by no means the only instance in history, nor the last in our story, where the mathematicians, with their uncanny instinct, had anticipated the future mathematical needs of science. Outside our story, the most famous anticipation of this sort was the tensor calculus of the Italian geometer M. M. G. Ricci which, when the time came, furnished Einstein with just the tool he needed for the development of his general relativity theory of gravitation.

Although Heisenberg had nurtured his calculus within atomic theory, it was not yet a theory of the atom. In one sense it was far more, being a new philosophy for science, but this was found out only later. Meanwhile it was little more than a calculus and a suggestion as to its uses. Out of it Born and Jordan undertook to create a new theory of the atom, and, more than that, a new science of mechanics— matrix mechanics. If p times q was not the same as q times p, then Born and Jordan must somehow discover what was the difference between them. The new matrix ideas were incompatible with the Bohr theory. But where else could Born and Jordan turn for inspiration? No other atomic theory was available. The correspondence principle had served to bridge the gulf between the Bohr atom and the classical mechanics of Newton. It must now be made to bring these two within reach of the matrices. The connection was the slenderest, but none the less suggestive. In it Born and Jordan found the needed clue, and from the old $\oint pdq = nh$, with much extraneous assumption, they finally extracted the following momentous equation:

$$p \times q - q \times p = \frac{h}{2\pi\sqrt{-1}}.$$

What this equation asserts is even more startling than the initial discovery that p times q and q times p were different. It states that their difference is equal to Planck's constant h divided by twice π times the square root of minus one. The square root of minus one is not an arithmetical number at all. Mathematicians sometimes call it an imaginary number because no "real" number when multiplied by itself can give the result minus one; two minuses give a plus.

That such a formula should have any connection with that world of strict experiment which is the world of physics is in itself difficult enough to believe. That it was to be the deep foundation of the new physics, and that it should actually probe more profoundly than anything before toward the very core of science and metaphysics is as incredible as must once have seemed the doctrine that the earth is round.

It was now September of 1925. More remained to be done before a new mechanics of the atom was created. Born and Jordan had already made a tentative advance toward fusing the new ideas with those of Newton's classical mechanics. Now Born, Heisenberg, and Jordan were to pool their very considerable abilities in a determined attack on this recondite problem. By November they had made sufficient progress to warrant publication of their researches.

But by now another youngster had entered the field; an Englishman named Dirac. Practically the same age as Heisenberg, he was able to do with ease and grace what the combined talents of Born, Heisenberg, and Jordan could accomplish only piecemeal and with considerable labor. When these three were pursuing their joint research, Dirac attacked the problem independently with a new idea that lightly brushed aside the formidable difficulties being encountered by Born, Heisenberg, and Jordan. And when, in January of 1926, Pauli at last succeeded in proving the crucial fact that Heisenberg's new theory would correctly yield the Balmer ladder of hydrogen, it was the same Dirac who announced a highly abstract generalization of the Heisenberg theory, and applied it to obtain a somewhat simpler derivation of the Balmer frequencies.

But so important a man as Dirac should not be introduced at the tail end of a chapter on Heisenberg. His place is in a chapter of his own.

CHAPTER X

THE ASCETICISM OF PAUL

PAUL ADRIEN MAURICE DIRAC tried to become an electrical engineer but, fearing he might not have the aptitude, turned his attention to the abstract physics he found more interesting. Whether he would have made a successful electrical engineer despite his misgivings is fortunately an academic question, for in his early thirties he was to be elected, appropriately, to that professorship at Cambridge University which had once been held by the great Isaac Newton himself.

Though Dirac merits an individual chapter, his chapter must be brief out of all proportion to his subsequent importance, a mere prelude to his later appearances in our story. For if Heisenberg's concept was so abstruse as to seem almost devoid of pictorial significance, the contribution of Dirac in the fall and winter of 1925 was the quintessence of abstraction, impossible to visualize, apparently, outside its mathematical context. But that was because it was then still a fledgling theory. Later we shall see that, for all its abstraction, which the passage of time was to increase rather than diminish, it could be readily and significantly visualized. With this promise of clarity to come, let us here take note, however sketchily, of the sort of ideas Dirac was already thinking in those early days. Though our outline must be sketchy, it may at least indicate the peculiarly astringent flavor of Dirac's early discoveries.

The announcement of Heisenberg's theory struck immediate fire in the mind of Dirac. Independently of the researches of Born, Heisenberg, and Jordan even then under way, he undertook to create out of Heisenberg's idea a new theory of mechanics. If x times y was not the same as y times x, then Dirac must somehow discover what was the difference between them, and using the indispensable correspondence principle he sought an analogue in the classical mechanics. In the classical theory there existed certain mathematical quantities, denoted by the symbol [x,y], which, having been discovered by the Frenchman Poisson, were known as Poisson brackets. Dirac, to his intense joy, discovered a relationship of extraordinary simplicity: calculate the value of the Poisson bracket [x,y] according to the classical theory, multiply by Planck's constant and the square root of minus one, and divide by twice π. Then the result will be the proper value to assign to the difference between x times y and y times x.

Does this perhaps seem a rather arid discovery? Dirac once said the most exciting moment of his life was the moment of its revelation. In one swift, dazzling leap, Dirac had surmounted the innumerable obstacles and difficulties impeding Born, Heisenberg, and Jordan in their efforts to fashion the new matrix mechanics in the image of the classical mechanics, and actually published his results a little before they could publish their equivalent, though less elegant, discoveries.

Dirac's initial discovery led him further, along a path of deep abstraction. Contemplating Heisenberg's theory, he now realized that its emphasis was misplaced, that it hid the forest with the trees. Although the huge square tabulations had been Heisenberg's chief inspiration, Dirac pointed out in January of 1926 that they were really incidental; they had no place in the central core of the theory but were outgrowths of something more fundamental. Stripping away the scaffolding which Heisenberg, and Born and Jordan had mistaken for the building, he fixed his gaze upon the strong, slender edifice beneath. As the Curies extracted a minute speck of radium from a mountain of ore, so did Dirac distill from Heisenberg's enormous square tabulations their ultimate essence, their one essential concept, that x times y may differ from y times x.

Science must henceforth be prepared to deal with two different types of "numbers," said Dirac. Along with the ordinary numbers it must use what he termed "q numbers" defying the ordinary rule of multiplication that x times y is equal to y times x. The p's and q's of the classical mechanics, which classically were ordinary numbers, must now be regarded as q numbers, and the new science of quantum mechanics, as distinguished from matrix mechanics, must be created out of them.

But without the square tabulations is there much left of the theory? Has Dirac really gone beyond Heisenberg? Does it not seem rather that he has not gone so far? Let us bear with him just a little longer.

The p's and q's of classical mechanics, then, are to be regarded as q numbers. So are the energy and the time, and all other such dynamical quantities which Heisenberg had conceived to be vast square tabulations. What else? Why, in a sense, *nothing* else! That was Dirac's great discovery. The classical mechanics could be made over into quantum mechanics by this one device, by means of the Poisson brackets. For the Poisson brackets were powerful entities in the classical theory, capable of representing the basic classical equations in simple form. To create the new quantum mechanics, one could write these equations of the classical theory absolutely unaltered, and merely reinterpret the Poisson brackets according to Dirac's earlier prescription.

Matrices? They were secondary. By a simple operation there could be generated from Dirac's equations the identical square tabulations with which Heisenberg had first explored the new world that science was entering. The new mechanics of the atom, for all its youthful radicalism, was shown to be a legitimate and fitting heir to the great and honorable tradition of classical mechanics.

Bohr had devised his correspondence principle in desperate appeal to the classical theory for aid. Heisenberg and Dirac had found its deeper significance. This was its culmination— a profound and abiding relationship between classical and quantum mechanics.

CHAPTER XI

ELECTRONS ARE SMEARED

OUR story is far from told. Even as physicists were frantically exploring the untold riches hidden within the Heisenberg theory, Einstein's forthright commendation of de Broglie's ideas was revealed as a major factor in the evolution of physics. For, toward the end of 1925, Einstein's words of praise brought the still-unconfirmed ideas of de Broglie to the attention of the Viennese physicist Erwin Schrödinger, at the famous University of Zurich in Switzerland.

The effect was galvanic. Within a few short months Schrödinger produced singlehanded a successful theory of the atom, only remotely related to the idea of de Broglie and utterly distinct from the theories of Heisenberg and Dirac. Nor was there anything strange about the mathematical methods of the new theory. So familiar were they, in fact, that even in his first announcement Schrödinger was able to carry through the solution of the basic problem of deriving the frequencies of the normal hydrogen atom, the problem that had so sorely taxed the skills of the Heisenberg group. The solution was sent to the publishers in January of 1926—the month in which Pauli and Dirac had independently sent in their own solutions of this selfsame problem. No extraordinary astrological skill was needed to descry in this a portent of boisterous happenings in the world of physics.[1]

Schrödinger adopted a curious method of announcing his theory to the world. He neither explained how it grew in his mind, nor indicated a complete logical sequence of ideas. He merely reminded his readers that a certain well-known mathematical process yields series of numbers which might be used as quantum numbers, abruptly wrote down a so-called wave equation—now known as the Schrödinger equation—and proceeded forthwith to extract from it a magnificent solution of the crucial hydrogen problem. This caused a startled outcry from the world of physics. Scientists are not interested in such displays of legerdemain. They want to know how and why the tricks work. Not enough to present them with a fait accompli. They want to know what lies behind it. Impressed by the distress of his fellow physicists, Schrödinger disclosed the secret of his sorcery, explaining in his second article how his theory was a natural extension of the ideas of de Broglie, and of the classical mechanics of Newton as developed by that outstanding Irish genius William Rowan Hamilton.

It is high time indeed that the name of Hamilton should be brought into this chronicle of events. For, though he died in 1865, his work was a dominating influence not only in the theory of Schrödinger, but also in the theory of Heisenberg before that; and before that, in the theory of Bohr; and even before that, in the theory of Planck. It was he, for instance,

[1] Do you like coincidences? Schrödinger and Pauli both came from Vienna. The former was born in 1887 and the latter in 1900. Where is the coincidence in that? Why, 1887 was the date of the experiment by Hertz, and 1900 that of the discovery by Planck. Add to this the Bohr-Balmer coincidence and we have quite a trilogy.

who first showed the importance of the p's and q's in classical mechanics. Without his researches the quantum theory of today would have been seriously delayed. And had he lived to learn of the revival of the wave-particle conflict he would certainly have anticipated the modern developments—so close had he actually come to them himself.

The keynote of Schrödinger's first article was the existence of simple quantum numbers hidden amid the complexities of atomic spectra. Bohr had simply injected these quantum numbers into his theory from the outside; they are such things as the n of the f formula. Schrödinger wished to avoid such an artifice. A good mathematical theory of the atom, he felt, must use a mathematical method generating quantum numbers in a natural manner from within itself. Hunt the method, then, and let the physical meaning take care of itself.

It was more than five hundred years before the Christian era that the Greek philosopher Pythagoras discovered a remarkable relationship between music and number. If a plucked string gives forth the note C, a similar string of half the length will sound the C an octave above. A string one-third the original length will give the G above that; a quarter the length, C above that; a fifth the length, E above that, and so on. So delighted and inflamed was Pythagoras by his discovery that he decided there and then that numbers, wonderful whole numbers, must be the key to the universe; only to have his high hopes dashed to the ground by his other great discovery, the well-known theorem about the hypotenuse. For this theorem showed that numbers existed, such as the square root of two, which defied rational expression in terms of the whole numbers.

Nowadays we know that the original string was sounding all the different tones at once. Usually the lowest alone was loud, the others merely adding color, or timbre, to the tone. They were the harmonics we have already encountered. Thus a vibrating string actually contains within itself sequences of numbers such as Schrödinger sought.

Now, a violin string is not free to vibrate in any manner it may wish. Since its ends are secured, it may vibrate only in such a way that its ends do not move. This is, of course, an obvious remark. But it is also a cogent one. For it is just this fact which limits the vibrations and introduces the sequences of whole numbers. The string can vibrate as a whole like this:

or in two parts like this:

or in three parts like this:

or in four, or five, or six, or any other whole number of equal parts. But it cannot vibrate in two and a half parts like this:

for then only one end at most could remain fixed. Thus it is the obvious remark that the ends must remain fixed which is the crucial remark that brings in the sequence of whole numbers 1,

2, 3, . . .; and brings them in in the most natural manner possible, as is evidenced by the fact that the remark at first seemed so obvious as to be hardly worth mentioning.

Here, surely, is the strongest possible hint. Do we need further urging before rushing to apply this principle to the atom? De Broglie has already told us there are waves there for us to use. Never mind this fellow Heisenberg. Who knows whether his theory is really any better than Bohr's, or even as good. Here is a simply wonderful idea begging to be applied.

But wait. It is one thing to have a wonderful idea; quite another to see how to carry it through.

What? We want further assurance? We are still timid? We still hesitate to take the plunge even after all this? De Broglie has yet another inducement for us, an almost irresistible one. Consider the evidence of a slender steel ring. When such a ring is struck it vibrates musically. We cannot say it does so because its ends are fixed. But, being circular, its vibrations are just as surely limited, and in much the same manner, for it is as if it had two ends that were joined together. It can vibrate as a whole, or in two parts, or four, or six, but not in two and a half. It can vibrate like this:

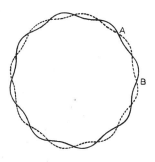

But it cannot vibrate like this:

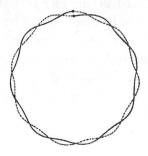

for here the wave after circling the ring does not join up with itself. Since a wavelength is the distance from A to B in the diagram, there must always be a whole number of wavelengths in the ring. Does this not remind us of something? Something about a length going an exact number of times around a circle? Why, yes, of course! The Bohr orbit condition. The strange business of the trolley segments that might not be broken up but must fit the orbit exactly. It always had seemed arbitrary and artificial. If we now calculate the de Broglie wavelength of the wave accompanying an electron in a Bohr orbit we find a truly fascinating result—the wavelength comes out to be precisely the special length of the track segment for the orbit. That makes the whole picture clear. It is the steel ring all over again, as de Broglie had actually pointed out in 1924. Almost the atom is beginning to make sense.

But where was Schrödinger to start? De Broglie had had the idea of waves accompanying electrons for some time without being able to construct a theory of the atom. Perhaps that was because he was thinking in terms of relativity. There was some highly suggestive work by Hamilton, done long before relativity, which seemed to fit in somehow with all this. Perhaps by leaving out the relativity one could progress faster. Clearly, what was needed was a wave equation. The whole history of wave motion pointed to this. It had long been known that the vibrations of strings and organ pipes, of kettledrums, jellies, and light waves, were governed by wave equations of similar types. It was known too that a wave equation would generate sequences of numbers as soon as extra mathematical conditions were imposed. And these were natural conditions. They said, in mathematical language, that the ends of the string were stationary, that the rim of the parchment of the kettledrum was secured, and reasonable, pictorial things like that.

Schrödinger decided to create an atomic theory out of such ideas. Concealing the secret of his manipulations, he made a few mathematical passes, uttered a judicious selection of mathematical invocations and incantations, such as *Hamilton's Partial Differential Equation, Minimal Integrals,* and *Quadratic Forms in Phase Space,* and magically produced, as if from nowhere, a full-grown wave equation having remarkable powers.

It was not a wave equation applied to a string, nor yet one applied to a membrane, but one applied to an essence filling a mathematical fiction of a space of a sort well known to mathematicians. The essence was represented by the Greek letter psi, ψ.

Schrödinger's ψ-essence—if we pause here to try to think of its meaning we are lost—was free to vibrate as it pleased with one proviso. It was, mathematically, fastened down at the uttermost bounds of the fictional space. It was this which was to bring in the quantum numbers.

Now, of course, one would expect the story to go on to tell that when Schrödinger applied his wave equation to the problem of the hydrogen atom he triumphantly found that the frequencies of his ψ-essence were precisely the frequencies in that atom's trademark, the frequencies in its spectrum.

But this was not the case. Schrödinger's triumph was of a somewhat embarrassing sort. The frequencies of the ψ-essence turned out to be those which belonged to the rungs of the Balmer ladder. Since only the differences of these frequencies appeared in the spectrum, this posed a pretty problem. Gone were the former electrons and their orbits. They had been swallowed up by the new ψ-essence, a vibrant smear of electron surrounding the nucleus. Without electron jumps, how explain the differences of the frequencies?

Schrödinger had a plausible explanation, as plausibility went in atomic physics. Was there not a similar thing in music? When two notes are not quite in tune, is there not a beat note?

If one note vibrates a hundred times a second and another a hundred and one, and both begin in step, they will be diametrically out of step a half second later, for one will have performed fifty complete vibrations and the other fifty and a half. As with the millionaires, their effects will then be nullified. But half a second later they are back in step again reinforcing each other, and combine to give their maximum effect. This rhythmic alternation from cancellation, through reinforcement, to cancellation again continues at the rate of once per second—the difference of the parent frequencies. The difference frequency is born of the marriage of the two original frequencies—even in physics marriage begets differences.

The new frequency is called the beat frequency and may be plainly heard as a beat or throb when two people whistle almost the same note. The squeals and howls in radios result from beat frequencies. The beat has even been exploited commercially by organ builders, who create certain tremolo effects with two pipes kept purposely out of tune so that the beat will produce the desired throbbing.

It was now Schrödinger's turn to exploit the beat frequency. His atom vibrated with frequencies which belonged to the rungs of the Balmer ladder. The frequencies required, being their differences, were none other than the various beat frequencies. Now, the ψ-essence was an essence of smeared-out electron, and a vibrant electron was known to give off light, even though Bohr had not hesitated to negate this to suit his special purpose. Let it be dogmatically asserted, then, that the beat frequencies of the ψ-essence were converted into light and we have at once the explanation of the atom's trademark, especially if we ignore the many objections one might raise. Schrödinger himself later offered a somewhat different picture, but it need not concern us here.

The important thing was the emergence of the required frequencies from the mathematical calculations. If the physical picture that went with these calculations was still somewhat obscure, the same could surely be said of the theories of Heisenberg and Dirac, and even of that of Bohr. Enough that the answers were correct. The door into the unknown had been wedged open a tiny crack. Science could now apply pressure and swarm through to the other side, and time would take care of the present obscurities. The way ahead was now well indicated.

Too well indicated, in fact. The problem of the Balmer ladder, which once had seemed insoluble, had now been solved in at least three different ways. Faced by the profusion and peculiar interrelations of the spectral frequencies, Bohr had conceived of electrons jumping from orbit to orbit. Heisenberg had attacked the problem by replacing laundry lists with square tabulations. Now Schrödinger had found a third interpretation. For him it had meant neither electron jumps nor square tabulations, but the beats of a vibrating electrical essence ψ.

Is it possible to have too much success? Physicists were sorely tempted to think so. Where in 1912, little more than a dozen years before, there had been no competent theory of the hydrogen atom at all, now, in January of 1926, there were as many as four, if we count those of Dirac and Heisenberg as distinct. Had they all betrayed some hint of a family likeness it might have been less perplexing, but what possible resemblance could one detect between the Bohr orbits, the Heisenberg tabulations, and the Schrödinger ψ-essence? Here was too bewildering a profusion.

What were physicists to think, where should they turn, that turbulent January? Bohr's theory, wise and experienced, could point to many victories in the past, but now it was old and suffered severe disabilities. The more youthful theories, relatively untried, could point to fewer triumphs, but already they gave promise of outdoing the theory of Bohr, for they proved immune to the ills that afflicted the Bohr quantum numbers, nor had they yet suffered any defeats. Matrices and waves were running neck and neck, with neither able to show a decisive advantage. Heisenberg's early triumph with the oscillating particles in July of 1925 was to be duplicated by Schrödinger in February of 1926. Dirac had shown that Heisenberg's theory was of noble birth. But, even as the race was on, Schrödinger revealed that his theory too was a thoroughbred, directly descended from de Broglie and Hamilton. Here is the far from dishonorable secret of its birth:

Newton's mechanics was built on his three laws of motion. But beneath these laws lay a deep foundation of numerous fundamental concepts which, though once revolutionary, came to be so unthinkingly taken for granted that Einstein's relativity amendments to them at first seemed highly unnatural. These underlying concepts, the Newtonian philosophy of space and time and matter, were essential preliminary assumptions without which the laws of motion could not be formulated, nor mathematics take hold to convert them into equations. When Lagrange and Hamilton made their great contributions to the development of Newtonian mechanics they did not call its philosophy into question, for in those days it was not fashionable to tamper with fundamentals; the aim was rather to develop them to their mathematical utmost in the sure belief they would then explain the universe.

Newton's equations of motion told the way in which bodies moved. A stone thrown in the air presents a simple enough problem, but it is fatally easy to cite more complex examples. Suppose we took odds and ends of machinery from a junk pile, joined them together with springs and elastic, and heaved the whole wobbly mass into the air with a sudden, vicious twist. Newton's laws would still apply, in theory, but the ensuing motions would be far too intricate for mathematical comfort. Though various methods were discovered for reducing the mathematical complexity of such problems, it was not till a hundred years after Newton that the great French mathematician J. L. Lagrange achieved a really notable simplification; but when the simplification came it was notable indeed.

One aspect of it especially claims our attention. The heap

of junk we threw into the air has so many motions going on at once that it is next to impossible to see what is happening. By a simple mathematical trick, however, we may remove this complication. We cannot completely destroy it, mind, but we can diminish it and put it where it is less obtrusive. The trick is to invent a fictional space having so many dimensions that the whole complex motion of the junk heap may be indicated by the tortuous movement of a single indicator point within it.

Let us not shrink from such a fanciful conception. We ourselves habitually employ fictional spaces in our everyday lives, for what is a hospital chart of a patient's temperature but the trace of an indicator point in a fictional space having two dimensions? Most of the times we draw a graph we use a fictional space. Here in dynamics, to be sure, the fictional space has many dimensions, but that is the whole secret of the trick. The complexity is now hidden away among the numerous dimensions of the space, and the bewildering motion of the junk heap is reduced to the easily imagined motion of a single point. True, the point moves in a complicated space, but the concept of a moving point is far simpler than the motion of the flying junk heap it represents. And being far easier to think about, it stimulates further discovery.

The further discovery was made by Hamilton, a man of superlative intellectual gifts. Hamilton it was who, even before Cayley and his matrices, and in a different connection, discovered that there are nonarithmetical quantities such that x times y need not be the same as y times x. Hamilton it was, too, who made over the fundamental equations of dynamics into the simple form, involving the p's and q's, which was to be the basis of all subsequent theoretical researches in atomic physics, and was to supply those researches with their central mathematical concept, an expression for energy now known as the Hamiltonian function. At thirteen years of age Hamilton had mastered thirteen languages. At twenty-two he was already a professor. And in 1834, at the age of twenty-eight, he had transformed the science of mechanics in the manner now to be told, and thereby all but anticipated Schrödinger.

However contorted the path of the indicator point in its fictional space, Hamilton knew he could bend a fictional light ray to fit it, for a light ray does not have to travel a straight line. A simple prism will bend it, or a lens, or even heated air. Almost any lack of uniformity where it travels will make it deviate. The desert air is hotter near the sand than higher up, and this inhomogeneity, bending the rays of light, causes mirages. What motorist has not observed on a hot dry road a fleeting shimmer as of cool, rippling water? This too is a mirage, ephemeral witness to the light ray's curvature.

By choosing the right sort of inhomogeneity, Hamilton could duplicate the path of the indicator point with his ray of light and thus forge a link between the sciences of light rays and dynamics. This being so, it must logically follow that these two different sciences are mathematically identical. Hamilton succeeded in proving this in rigorous mathematical detail, all, of course, in terms of the fictional space. But does that not of itself bear witness to the fruitfulness of introducing that space? Could one have imagined the optical connection in terms of the flying junk heap? (An optical connection with that would be quite painful.)

Hamilton did more than duplicate the path of the indicator point with a ray of light. A light ray corresponds more or less to the path of a *particle* of light. Hamilton went further than the ray or particle by introducing part of the concept of the wave, that part of it which controlled the path of the ray. Thus Hamilton had already reduced the whole science of dynamics to the study of waves of light, but waves of light lacking the crucial millionaire property, the property of interference. Now, in the science of optics, rays are quite adequate for investigating the simpler properties of optical instruments, but when minute effects are to be explained interference of waves must be invoked. How if the same were true of classical mechanics? Just as rays of light sufficed for calculating a pair of prism binoculars, Newton's dynamics was quite adequate for large-scale phenomena, but broke down for systems of atomic dimensions. How if the truth should be that Newton's was a "ray" dynamics, while what was needed for the minute dimensions of the atom was a wave dynamics? It was an alluring possibility.

Something very like it had already been indicated by de Broglie. De Broglie had been thinking of the actual space and time of relativity. But here, long existent and already highly developed, was an optical dynamics in a fictional space, which lacked but one small ingredient. Schrödinger, realizing the technical difficulty of working with relativity, saw the great possibilities in Hamilton's optical ideas. Taking Hamilton's incomplete waves, he endowed them with the property they lacked, the power of interference. Now they were true waves, yielding the same rays as before for large-scale phenomena, but capable of exhibiting entirely new properties when applied in the realm of the atom.

How natural a step this was can be seen from the history of optics, for optics too had begun as a science of rays only to find need for wave properties when faced with more refined optical phenomena. How successful a step it was is attested by the history of the dazzling months that followed Schrödinger's initial announcement.

Very early, the Schrödinger theory was a strong competitor of the Heisenberg theory as to results, and bade fair to outstrip it in general popularity. It avoided the formidable technical difficulties of Heisenberg's theory. It offered a comforting picture of atomic processes. Its noble lineage could match that discovered by Dirac, and its warm, pictorial character carried appeal. It could produce its results with comparative ease, speaking the mathematical language of the ordinary theoretical physicist. It did not require him to delve into unfamiliar mathematics, nor to invent special methods for each new problem. By yet another of those uncanny anticipations which threaten to mar the artistry of this story, its mathematical methods were already prepared for it, and neatly packaged awaiting its arrival. Two eminent German mathematicians, R. Courant and D. Hilbert, not physicists, but leaders of the Göttingen group of mathematicians, had written a book called *Methods of Mathematical Physics*. This book, in compact, convenient form, contained practically every mathematical method, trick, device, and special detail required for the development of the Schrödinger theory, not to mention much that was applicable to the theory of Heisenberg. The date of its publication was 1924.

UNIFICATION

A PRETTY piece of juggling science does here! The Bohr theory has fallen apart in its hands. But, for the moment, it contrives to save the act by tossing into the air, in quick succession, the two pairs of theories of de Broglie, Heisenberg, Dirac, and Schrödinger. Now it juggles more merrily than ever. Science was surely not meant to play the juggler. Yet here are four dazzling theories in the air at once, and so far none has fallen to the earth.

Four theories, in fact, very much in the air, and so far none has really come down to earth.

But science will not continue long like this. Its internal sickness is at a crisis. The fateful moment has come. Turmoil and ferment have reached their distressing climax. No further theories are destined to arise to confound confusion more. For ailing science has fashioned for itself potent new physics. Gentle healing will soothe its troubled frame, and bring it strength far greater than before. Wave and particle will be reconciled. Divergent theories will come together, and with their meeting will come new understanding. This has all been the agony of travail out of which will be born a greater, and more humble, science.

Already a certain order is apparent. Clearly the theories of de Broglie and Schrödinger are similar, as too are those of Heisenberg and Dirac. Thus there are only two main lines of progress. But these are so unlike that any hope of a rapprochement must seem vain.

Yet something is suspect about this dissimilarity. Neither theory has established a clear-cut advantage over the other. Indeed, each seems to ape the other's triumphs. There is, too, the mystery of their heredity. Each boasts of its exalted lineage, claiming to be the only natural heir of classical mechanics. Yet one is descended from mechanics and the other more from optics. But did not Hamilton himself link optics with mechanics in the classical theory? Why, then, should not the two new theories be brothers, maybe twins, beneath their surface differences? Their seemingly divergent origins may well be one. It is not natural that two such plausible offshoots of classical mechanics, neither one able to out achieve the other, should be so distinct as their outward appearances pretend. It is not natural that two really different theories should long wage war over the same group of facts.

But what of the wave and particle? Are they not warring still?

Perhaps there is a hidden unity beneath our story. Perhaps this battle of the theories is nothing new. Let us examine it a little more closely. The ideas of Heisenberg and Dirac stem from the particle dynamics of Hamilton, those of Schrödinger from Hamilton's wave dynamics. This Heisenberg-Schrödinger controversy may thus be but an extension or reflection of the ancient strife between particle and wave. Wherever we turn, that battle intrudes as a central feature of science. But Hamilton himself now holds out promise that the two may be reconciled. Perhaps in bringing the theories of Schrödinger and Heisenberg together in common birth he brings them together too with the wave and particle.

As early as March of 1926, a brief three months after the appearance of his theory, Schrödinger took a decisive step toward unity. Once again the secret lies with Hamilton, whose creation of optical dynamics was more than the pointing up of an analogy. The ordinary dynamics required many equations of motion. But Hamilton could write a single equation to govern his pseudo waves, just as a single wave equation governs genuine waves. Thus Hamilton could now reduce the whole science of classical dynamics to a single equation, truly a momentous and monumental achievement.

Schrödinger had endowed Hamilton's pseudo waves with the power of interference. Surely, then, there must be some connection between Hamilton's equation and the wave equation of Schrödinger.

Yes, to be sure, there was a vague sort of connection. Enough of a connection to still further inquiry. But one day there flashed upon Schrödinger a far deeper relationship, a relationship that was enormously exciting. He found he could convert the first equation into the second by a superbly simple mathematical trick. Wherever p occurred in Hamilton's equation it must be replaced by a certain mathematical entity called an "operator." Never mind what the precise operator was. The important thing was that the step from classical to quantum mechanics could be made by replacing p by an operator. You want to see the operator? It really is not necessary. You want to see if it is pretty, like that Bohr orbit thing? Yes, it is pretty. Take a look:

$$\frac{h}{2\pi\sqrt{-1}} \quad \frac{\partial}{\partial q}$$

There is that h again, and that square root of minus one, and that 2π. They certainly stick together.

In mathematics an operator is not a number but a command. It is an order to perform some particular mathematical operation. For example, "multiply by 2" is an operator, so is "add 3." The mathematician would write these operators more compactly, but he would still mean these same commands. When two operators are intended to be applied in succession they are said to be multiplied together. This agrees with our ordinary ideas in the cases of such operators as "multiply by 2" and "multiply by 3," for applying these in succession is the same as multiplying by their product, 6. Let us give here a simple illustration of the potency of operators. Since the operation of multiplying by minus one reverses the sign of a mathematical quantity, it may be aptly regarded as corresponding to the military command "about-face." What command, then, would correspond to multiplication by that so-called imaginary quantity, the square root of minus one? Nothing very mysterious, as it happens. In fact, it is something quite prosaic, for it must be a command which on being fulfilled twice results in an about-face; that is, the command "right turn," or else the command "left turn." Even the ambiguity is appropriate, for it is well known that a square root has an ambiguous sign. This particular representation of the square root of minus one is widely used in mathematics. Simple though it may seem—indeed, because of its very simplicity—it has exerted a profound influence on the course of mathematical thought.

There is a significant property of operators which comes in

most aptly at this point. Suppose we apply the operators "multiply by 2" and "add 3" in succession to the number 1. Multiply by 2 and we get 2. Add three and we get 5. But now let us apply them in the opposite order. Add 3 to 1 and we get 4. Multiply by 2 and we get 8—*which is not the same answer as before.*

This, of course, is of the utmost importance. If p and q are operators, then p times q and q times p need not be equal. Schrödinger's discovery goes even further than this. When the difference between his operator p times q and his operator q times p is calculated, and it is a calculation a beginner in the calculus could perform, the result, for whatever the thing operated on, turns out to be always the same—precisely the quantity found in the Heisenberg theory. What Schrödinger has done is to make Dirac's discovery all over again, but in terms of waves instead of particles. Clearly it was the identical discovery, underneath. But it was even more than that. It demonstrated that the whole Dirac theory of q numbers was implicit in the Schrödinger theory of waves.

So much for Dirac and his q numbers. But what of Heisenberg and his matrices? Schrödinger had by no means finished yet. By a none too intricate mathematical process, making liberal use of his ψ, he showed how the p's and q's and similar quantities in his theory could be dissected, and their bones laid bare and neatly displayed for all to see. And when they were properly arranged these bones filled vast square tabulations; they were precisely the Heisenberg matrices.

Now Heisenberg's theory too was contained in Schrödinger's. We could even have guessed as much as soon as Dirac's theory was swallowed up by it, for Dirac himself had shown that Heisenberg's matrices were latent in his q numbers.

How vastly changed is the picture of theoretical physics but three short months after its tangled hour of crisis. Schrödinger's theory, with its familiar picture of waves, far easier to manipulate mathematically than the theories of Heisenberg and Dirac, has now completely gobbled up its rivals. Those theories were its skeleton, so to speak. No wonder the Schrödinger theory was easier to visualize. Who would now want to go back to the matrices or the q numbers? They were but fossils, evidences of the intermediary stages in the evolution of the quantum. Now the theory at last is fully revealed. Schrödinger is the victor, and all is well. Here is the ideal place to end the chapter. An era of tumult has ended. Peace is at last at hand.

But no! The story of the quantum is not so simple as this. The last paragraph is sadly mistaken. It is premature in its jubilation. The chapter must go on.

There is a different aspect of the situation. Was not Schrödinger's discovery almost as much a vindication of Dirac's theory as of his own? True, Schrödinger possessed the ψ, which Dirac lacked. But Dirac had insisted all along that the great square tabulations of Heisenberg were only secondary, and Schrödinger had found the strongest possible corroboration since his operators in no wise resembled the Heisenberg matrices. When clothed with the flesh and blood of his ψ they were revealed as quite simple, familiar operators of the calculus. Only after quite detailed mathematical dissection could Schrödinger lay bare their Heisenbergian skeletons. Schrödinger's theory may have gobbled up Heisenberg's, but in so doing it had but vindicated the early intuition of Dirac. Soon it was to be Dirac's turn.

The months following Schrödinger's discovery teem with activity. The new ideas leap swiftly from triumph to triumph. From all sides come reports of brilliant conquests, by wave and matrix and q number. The complexities of the Zeeman effect are quickly vanquished. The details of the Stark effect are explained in all their intricacy. By June, Heisenberg has found a brilliant explanation of hitherto baffling features of the spectrum of helium utterly beyond the powers of the old Bohr theory. Simultaneously Born announces a profound discovery, at last revealing the true meaning of Schrödinger's ψ. Even Heisenberg begins to use the ψ's.

In August, Dirac grafted a ψ on the bones of his q number theory, the q numbers proving ideally suited to receive the graft. To flex the new-found muscles of his theory, he showed how Schrödinger's sophisticated extraction of the Heisenberg matrices could now be performed in a really simple fashion. Not content with this small exercise, he followed up Heisenberg's ideas on the helium spectrum to give what remains to this day the nearest we have come to an explanation of the mysterious principle of Pauli which prevents electron overcrowding.

But all this was a preliminary trial of strength. The idea was destined to grow enormously in power. By December, Dirac had made over his q numbers and their borrowed ψ into what is still the most comprehensive and catholic formulation of the rules of this new game of quantum mechanics the theoretical physicists had discovered. Jordan discovered the rules independently about the same time.

To appreciate what was accomplished, let us consider the rules of that far more ancient game, chess. Twenty-five cents will buy a booklet expounding the rules of chess with the utmost lucidity; and the rules of checkers, halma, and dominoes, and of an enormous number of incomprehensible card games for good measure.

Everything about the rules of chess is told; the names of the pieces, how each one moves, such refinements as how to take a pawn en passant, and, with luck, even the simpler standard openings and end games. It is lucidity itself. What more could one want?

All is not truly perfect. Something has been put into the rules which does not belong to chess. The rules as given are adulterated. They are written in English. Naturally, that is no valid ground for demanding one's quarter back. But in principle it is a serious fault. What, after all, has chess to do with the English language, specifically? Is it not played all over the world? Is there a different chess in France? The Chinese play it, and so do the Russians. Show your twenty-five cent booklet to a native Chinese or Russian and the chances are he will be unimpressed with its much-vaunted lucidity. Most likely he will give you tit for tat by thrusting under your nose a neat booklet of his own, in which the rules of chess are exquisitely described in limpid Chinese or in Russian of crystal clarity.

To the average American, the rules of chess written in Russian will seem to have nothing in common with the same rules written in Chinese. But let him see a Russian and a Chinese play an actual game and the connection becomes immediately obvious.[1] Though the rules of chess look different in different languages, the game is as universal and as free from the trammels of language as music or toothache. Theoretically, the universal way to describe chess is to procure a board and set of men and proceed to demonstrate by means of sign

---

[1] Unfortunately, if the truth must be told, the Chinese game is slightly different from ours. But the Chinese language has such a picturesque appearance, let us not permit a mere fact to spoil our analogy. Facts may be stubborn things, as a certain Ulyanov once so trenchantly remarked, but surely they are not so stubborn as that.

language. How else could one who was not a gifted linguist explain chess to a polyglot group of emigrants on Ellis Island?

With any universal idea, the most natural method of description is often the most primitive and unsophisticated. Let a man be marooned on no matter what strange shore and, unless he has the misfortune to be eaten before he has the chance, he will always be able to convey to the natives that he is hungry or thirsty, or lacks sleep, or that he has a stomach-ache which is mild, or medium, or frightful. And all this with a precision and clarity of nuance such as the most experienced novelist finds hard to match in words. His signs can be rendered into any spoken language under the sun, producing different sounds in different languages. But beneath the confusion of tongues, this hunger, or thirst, or sleepiness, or particular degree of stomach-ache will remain, especially for the victim, the prime reality.

Dirac discovered what was the prime reality beneath the confusion of theories in the new quantum mechanics; the basic rules of the new game the physicists were playing. And he expressed them in the mathematical equivalent of sign language, primitive in form but amazingly precise in expression. Though these rules were extracted from the theories of Heisenberg and Schrödinger, they showed little trace of their origin. The q numbers were there, for Dirac had divined correctly from the start. So too was a ψ, but it was a far cry from the original ψ of Schrödinger. What became of that will be told in a moment.

Finding the fundamental laws of quantum mechanics in mathematical sign language was only part of Dirac's achievement. He also showed how to translate the rules into any mathematical language capable of expressing them; of which, for instance, arithmetic would not be one, any more than the language of the Australian aborigines would suffice for telling the story of the quantum. When Dirac wrote out the rules in one of these mathematical languages—let us call it mathematical Chinese—they became simply the theory of Heisenberg, with a ψ added. When, however, he wrote them in what we may call mathematical Russian, they became precisely the theory of Schrödinger. Dirac even constructed a universal "dictionary" for translating from any one mathematical language to any other. When he wrote out the special dictionary linking mathematical Russian with mathematical Chinese, that is, linking Schrödinger's theory with Heisenberg's, he found it consisted of none other than the ψ's of Schrödinger.

Such was the magnificent scope of Dirac's amalgamation. Schrödinger's theory had started the feast by gobbling up Heisenberg's and thinking it had gobbled up Dirac's. Now Dirac's theory had gobbled up everything, and those strange bedfellows Schrödinger and Dirac were to share the Nobel prize in 1933. Instead of becoming fatter and more slovenly, quantum mechanics had become successively more svelte and elegant. With Dirac's work, the main structural scheme of quantum mechanics was now established.

But what did it all really mean? What sort of mental picture could one form of it? For all its eloquence and incomparable achievement, it still remained somehow remote, obscure, and unfriendly.

Even while the above events were unfolding, Heisenberg was piercing the mists that still swirled about their theoretical foundations, and Bohr was soon to bring further enlightenment. What strange new realms of physics were thus revealed will be told in following chapters. Let us not pause for explanations here. This is a chapter telling of unification. And although the prime unification of them all has now been told, there is an urgency and momentum of events which sweeps us forward. There will be time enough for understanding. Let us pursue unification yet a while.

So far the new quantum theory had busied itself with matter, leaving light to fend for itself as best it could. And light's best was little better now than it had been under Planck and Einstein and Bohr. The theory of matter had burst suddenly into full flower but its consort, the theory of light, had lagged behind. Like a youth at puberty, it had remained in an awkward state of half-classical, half-quantum adolescence, and the slight, wavy down of photon which adorned its smooth, classical features deceived no one into thinking it had attained full quantumhood.

In the feverish atmosphere then prevailing, growth was rapid. In February of 1927 Dirac brought to the photon a swift maturity even more speedy than that which had so recently come to the particle.

Let us imagine a box lined with mirrors, top, bottom, and all around. Any light waves unfortunate enough to be trapped within it must spend their days in one mad, headlong rush back and forth in all directions, battering themselves repeatedly against the mirror walls only to be remorselessly and inevitably reflected back at every encounter.

Light waves will do curious things under such harsh conditions, as the English physicist James Jeans had discovered back in 1905 in an investigation connected with the violet catastrophe. Like madmen pretending to be Napoleon, light waves trapped in their mirrored cell will pretend to be a collection of oscillating particles; for Jeans showed that Maxwell's equations for light in a reflecting box can be so cunningly maltreated that, instead of looking like the usual wave equations, they will take on a remarkable resemblance to the ordinary mechanical equations of such oscillators—an infinite number of them, in fact.

It was on this discovery of Jeans that Dirac built his theory of light and its interaction with matter. Jeans had twisted Maxwell's wave equations into equations having p's and q's just as if they had been taken right out of Hamilton's mechanics. Here was a splendid opportunity. Clamping his own quantum ideas of q numbers on these p's and q's, Dirac converted this into a quantum theory of photons having far-reaching implications. Though the process may sound simple when stated baldly like this, it was an operation bristling with difficulties and demanding considerable virtuosity. New entities had to be introduced which, unlike such simple things as p's and q's, had no counterpart in the classical mechanics. New ideas of all sorts, both mathematical and physical, had to be kept on hand to plug the many leaks which threatened to bring the theory to grief. Not all the leaks were stopped by any means, yet, so long as he did not drive it too hard or too far, Dirac was able to keep the theory afloat.

Overnight Dirac had brought the laggard theory of light into the domain of the new quantum mechanics to be a worthy companion to the theory of matter. History had repeated itself. Here was the pattern of a decade before all over again. In 1917 it had been Einstein who brought the theory of the interplay of matter and radiation into line with the newly propounded Bohr theory. Now Dirac had performed a corresponding service for the new quantum mechanics. Of the general importance of

Dirac's theory of light in the quantum mechanical scheme, and the significant developments to which it has given rise, it would take us too far afield to tell. But there is one item which holds special interest as a crowning unification. For Dirac could now derive, with all the elaborate machinery of quantum mechanics and at last as an integral part of it, something which had hitherto remained outside: the various ideas Einstein introduced on general grounds ten years before, and the original empirical radiation formula of Max Planck which had started the whole thing off.

With Planck's immortal formula turning up for the third time, a veritable rock in a boiling sea, its form untouched by the passage of turbulent years, the quantum has now traveled full circle. Here is the place to terminate the chapter. Or is its momentum even now not fully spent?

There are still loose ends to be gathered together. What has been happening to de Broglie's waves all this time? What about relativity? And then, too, what of the spin of the electron, and that relativity formula of Sommerfeld's for the fine structure which got lost in the recent storm? How have all these fared?

One more amalgamation remains to be told, which knits all these together.

Where de Broglie had used relativistic waves in ordinary space and time, Schrödinger had used nonrelativistic waves in a fictional space. The extraordinary success of Schrödinger's theory soon made it seem that de Broglie's day was done. But without relativity or spin Sommerfeld's formula could not be resuscitated. Attempts of course were made to replace Schrödinger's waves by relativistic waves, but Sommerfeld's formula refused to show itself except in garbled form. Something was wrong with the theory. In this particular instance it was not even as good as Bohr's—fighting words by now.

Another puzzle had appeared meanwhile. If an electron was a wave, as Schrödinger said, how could one fit in the spin? This problem was attacked by Pauli, and independently by the English physicist C. G. Darwin, grandson of the Charles Darwin of *Natural Selection*. Pauli, following the Heisenberg tradition, sought to duplicate the effect of a spin by introducing special matrices, while Darwin, who felt more at home with Schrödinger's ideas, introduced a modified form of electron wave. By now it will occasion no surprise that the two theories were "Russian-Chinese" counterparts, as was shown by Jordan. When artificially combined with relativity, the new ideas brought Sommerfeld's formula back into the world of physics, except for a small discrepancy. But they suffered from a much more significant discrepancy than this, for they gave for a certain quantity exactly twice the value it was known to have from experiment.

It was at this point that Dirac, in 1928, took command of the situation by going right back to de Broglie and relativity, and leaving the spin to take care of itself. For Dirac, with his deep insight into the foundations of quantum mechanics, had noticed that de Broglie's simple wave equation must be regarded quantum mechanically as a two-ply affair. So cunningly did the two parts fit together, and so firmly were they bonded one to the other, that no one had hitherto suspected the duplex character. With great mathematical dexterity, Dirac pried the two parts asunder, and lo! each part was equipped with built-in matrices—built-in matrices exactly representing the electron spin. He demonstrated that either part alone was a sufficient wave equation for the electron and showed that his new equation not only brought back the Sommerfeld formula intact into quantum physics but even improved on it. It also removed the discrepancy of the value twice as large as it ought to be. It showed that the spin of the electron was but a natural reflection of relativity, thus resolving the apparent conflict of relativity and spin as to responsibility for the fine structure. On the mathematical side, it introduced new quantities in the theory of relativity leading to a new calculus called, in honor of the spin, the spinor calculus. It superseded the equation of Schrödinger for a single electron, it led to certain significant developments which will be told in a later chapter, and all in all it showed what wonderful results were to be obtained from a successful marriage of those two outstanding rebels of modern physics— the quantum theory and the theory of relativity.

Yet this successful alliance was more a marriage of convenience than a true union. For all its tantalizing brilliance, for all that it marked a profound penetration into the unknown, it did not bring relativity and quantum mechanics intimately together. Between the two there remained an element of incompatibility that seemed to cramp the activities of both so that, for instance, no proper way was found of applying the new equation to an atom having two or more electrons. Many problems arose in connection with it, problems made all the more acute by the dazzling success it had achieved. One of these problems, having to do with negative energies, is particularly fascinating. But it belongs to a later part of our story, for we have allowed the momentum of our tale to carry us farther than it should. By 1928 the quantum mechanical revolution was already over. Brave ideas and magnificent discoveries continued to arise, but the revolutionary rioting had subsided and quantum mechanics was already enthroned as ruler and leader of atomic science. Our story must now go back again in time to plunge once more into the thick of battle.

CHAPTER XIII

THE STRANGE DENOUEMENT

Macbeth: ". . . . . it is a tale
      Told by an idiot, full of sound and fury,
      Signifying nothing."
                    (*Macbeth*, Act V, Scene V)

Polonius: "Though this be madness, yet there is method in't."
                    (*Hamlet*, Act II, Scene II)

MACBETH OR POLONIUS, that is the question. Our tale has lacked nothing of sound and fury, or madness. Readers may even say it was told by an idiot. Yet there is method in it. It is not empty of meaning.

Understanding came late to the new quantum theory. Men carried the quantum forward to commanding heights without knowing what it signified. They worked aware that momentous events were abroad, but with as little foreknowledge of the meaning of their discoveries as a caterpillar might have of its destiny to become a butterfly. They had already scored spectacular triumphs when the first inklings of understanding began to appear.

Perhaps it is natural that understanding should be so long delayed, for the new concepts were strange and hard to accept.

They would have been rejected had they lacked their impressive array of corroborative evidence, an array that was well-nigh overwhelming. It was the unparalleled harmony between theory and experiment that forced the new ideas upon a none too willing science.

Early guesses, hopes, and mental pictures were to be discarded. What could have been more plausible than de Broglie's resolution of the wave-particle difficulty? He had the idea that his waves were an adjunct to the particle, not a substitute for it; that one could never have a particle without its attendant seeing-eye wave to guide it; to spy out the way ahead and nudge the hesitant particle along the one path which would require the least action. What could offer greater promise of solving the wave-particle mystery than such a wave-plus-particle concept as this? But it was not to survive. The mystery was to prove of greater subtlety.

Heisenberg had hoped to play safe by playing the ostrich. He had rejected all mental images and offered no unproved pictures of what might be going on within the atom, for it was such pictures, he felt, that had caused the Bohr theory's downfall. Dirac, too, began by renouncing pictorial imagery. He rushed heart and soul into his q number theory, seemingly undismayed that something might be lacking in the way of warmth and human good-fellowship.

Schrödinger snatched the de Broglie waves from their playground in space and time, removing them to the remoteness of fictional space and abandoning their former playmate, the particle. His electron was now smeared out, and lacked location even in fictional space. Let us tell a little incident in this connection. Noting the sorrow of his wave for its lost playmate, Schrödinger tried to give it location by piling up many small waves into one huge localized wave. He even proved mathematically that this "wave packet" was a good substitute for the particle, that it would not fall apart, but would move exactly as a particle would have according to classical mechanics. Unfortunately for this attempt to console the wave for its lack of location, it was later proved that, by a singular coincidence, Schrödinger had worked with the one type of problem where a wave packet would behave in this convenient manner. In almost all other instances the wave packet would fall apart. It would seem, therefore, that Schrödinger's idea for locating the electron was incorrect. But there is more to this than at present meets the eye, and we shall return to it later.

Many months after he introduced his theory, months during which he had been applying it with phenomenal success and even gobbling up rival theories, Schrödinger at last ventured on an interpretation of his ψ. It was to measure how thickly the electron was spread out, much as one might measure the uneven thickness of butter spread on bread. He gave a specific mathematical formula for it. The interpretation was quickly superseded. The actual formula survives.

In June of 1926 Born suggested that the electron was not smeared out after all, but that ψ is a measure of the probability of the electron's being in any particular place. We find a young American physicist discussing this concept. His name was J. R. Oppenheimer.

No sooner do we half reconcile ourselves to waves of a smeared-out electron than we are asked to replace them by waves of probability. How many more of these fanciful ideas must we hear before we get to the right one? And, at this rate, how are we going to recognize the right one when we do come to it, if ever?

We have just come to it! Schrödinger's waves are waves of probability. That, at least, is the accepted interpretation to this day, and there is nothing to indicate it is likely soon to be superseded. Indeed, it is quite fundamental for the interpretation of quantum mechanics, and is sustained by the strongest corroborative evidence. Even so it is a curious concept. Born must have found compelling reasons for adopting it. What can have induced him to abandon Schrödinger's idea of a smeared-out electron?

He was led to his new interpretation by considering what happens when, for instance, an electron nearly collides with a nucleus. If we treat the electron as a ψ wave we find that the ψ is splattered all over the place by the collision.

No great harm seems to have been done so far, though. Why should it not be splattered?

There is harm enough. The ψ, being a wave, can of course be splattered with perfect propriety. But what does that mean in terms of electrons? Does it mean the electron has been shattered into little pieces? Experiment is clear on that point. Electrons are not shattered in this way. They are deflected if they pass close to a nucleus. Yes. But they remain whole electrons. Yet Schrödinger's idea would imply that an electron could never survive a collision whole. It was an impossible situation for the smeared-out electron. The only way out seemed to be to regard the ψ as not so much describing the particular behavior of an individual electron as telling what the electron was liable to do on the average in very many collisions.

This is a difficult idea. Maybe we had better hurry on to

other ideas which make this new interpretation feel more reasonable. Until these other ideas came, the probability waves, and almost everything else connected with the interpretation of the new quantum mechanics, brought acute mental discomfort to all but a fortunate handful of physicists, and even those fortunate few, in the very forefront of the march of progress, breathed more easily when these other ideas appeared. We owe our understanding of the everyday meaning of the new quantum mechanics, and our ability to form consistent mental pictures of what goes on, in the first instance to the genius of Heisenberg; Heisenberg who began by renouncing all seductive mental images and hid them from his eyes lest they lead him astray. And we shall soon see how sound had been his instinct, for the new ideas revealed that any pictures he might have formed beforehand would surely have been grossly misleading.

Like everyone else who knew what was going on in theoretical physics, Heisenberg was puzzled and disquieted by the contrast between the clarity of the achievements of the new mathematical equations and the obscurities and uncertainties of their basic interpretation.

Once upon a time a baldheaded man asked a small boy whether he would not like to be bald too and have no hair to comb. With the quick apprehension of boyhood, the youngster replied, "Oh, no! That would make twice as much face to wash." That boy was well acquainted with the fundamental principle of natural perversity so familiar to all who wish to eat their cake and have it. Heisenberg was to discover a comparable perversity in the realm of physics. He began by asking himself some fundamental questions. It was all very well to say that p times q being different from q times p explained atomic phenomena; but it did not explain why p times q could not be equal to q times p. Of course, if we accepted p and q as operators we could see why they would be able to behave like that, but that was mathematics, it was still not an explanation. What about physics? What about experiment? After all, p and q were not just mathematical symbols. They were supposed to represent physical things: p was momentum, and q was position. What did the inequality of p times q and q times p mean in terms of actual position and momentum? What did it mean in terms of experiment?

That was the clue. What did it mean in terms of experiment? In 1927 Heisenberg found the answer, deducing it mathematically from the sign-language rules of quantum mechanics and clarifying it by many vivid physical illustrations. Dirac, independently, about this time, realized what was the true state of affairs, and Bohr was quick to grasp the deeper significance of the new ideas. Following Heisenberg and Bohr, let us see how graphically these ideas may be described. How does one make an experiment to measure p and q for some particle, say an electron?

That is easy. We simply look at it and note its position and its velocity. The position is q. And the velocity multiplied by the mass of the electron, which latter we know from other experiments, is p. How do we find its position? Just by looking at it, of course. How do we find its velocity? By looking twice, of course. We clock the electron much as we would a runner in a race. We look at it at the start and again after an interval of time, and note the change in position. What is all the fuss? It is all clear and aboveboard. It is a thoroughly routine procedure. Astronomers have been noting positions and velocities for centuries. Why not talk to them? They can tell how to go about it with almost unbelievable precision.

Yes, it has been done for centuries and centuries, but with planets, and stars, and nebulae, and asteroids, and satellites, and meteorites; and in mundane affairs, with trains, and airplanes, and swimmers, and race horses, and shellfire, and hurtling stones. But what about electrons? Electrons are small. They are very small. They are exceedingly small. To make the measurements we must be able to look at them. How are we going to see them?

Well, there is always the possibility of using a microscope.

But a microscope will not be anywhere near powerful enough.

We could imagine one powerful enough. What is the point of all this?

Even if we use a hypothetical microscope of simply phenomenal power we still have to look at the electron.

Naturally!

But looking at it implies shining a light on it.

Certainly! Of course! Everyone realizes that. No one is going to object. Let's get down to business.

We are getting down to business. There is a well-known rule about microscopes. Their powers are limited by the size of the light waves used. They cannot distinguish details smaller than a wavelength. Why do the most powerful optical microscopes use ultraviolet light, if not for that? Why are electron microscopes so much more powerful, if not for the fact that the de Broglie wavelength of fast-moving electrons is so much smaller?

It's a hypothetical microscope, anyway. Why not use hypothetical light with it? Use X rays if necessary, or γ rays from radium, or light of even shorter wavelength. Use whatever light is necessary, no matter how small its wavelength. It is all hypothetical. It does not cost us anything.

All right. We will use light of extremely small wavelength. But there is a well-known rule about light. The shorter the wavelength, the higher the frequency. And as Planck and Einstein discovered, the higher the frequency the larger the energy of the photon.

So we're using photons now. Is that really fair? We were merely talking about a microscope. No need to confuse the issue by bringing in photons.

But we *must* bring in photons if we want to think about the quantum theory. That is the fundamental point which was overlooked in all previous speculations of this sort. We know that light is somehow atomic, that each frequency comes in bundles of definite energy. How can we ignore so fundamental a fact if we want to understand the quantum? And what is all this commotion coming from our microscope now? The electron apparently doesn't like the new turn of events. It is having a rough time of it. We are not leaving it in peace any more. We are not just looking at it, we are hurling enormous boulders of energy at it and it is being badly knocked about. What sort of a scientific experiment is this? It is certainly far from delicate. Suppose we do manage to see the electron and note its position? It is an empty victory. The very fact that we see it means we have scored a direct hit with a photon. The electron is a very light particle, unable to withstand a particle of light. It is badly jolted by the impact. In observing the electron's position we give it a jolt which alters its velocity. We defeat our own object. We cannot use gentler photons, for the less their energy the less their frequency and the greater their wavelength, and thus the less the power of the microscope. A spirit of perversity is in the air.

But why not make the best of it by observing the original position, and then the velocity after the jolt of the initial observation?

That does not help at all. How do we measure the velocity? We have to use a mental stop watch and make two successive observations of position to see how the electron moved. The second observation would cause a second jolt. The velocity we so carefully calculated from our observations would not be the present velocity of the electron; the second jolt would have altered it at the very moment we completed our observations of it. We can know the past velocity, but not the present or future. The spirit of perversity is becoming positively obtrusive.

This is nonsense. It's ridiculous. It cannot be like that. There must be some way out. Why not take note of the jolt on the photon itself? Then we can calculate what jolt it must have given the electron, just as we could for a collision of billiard balls.

Very ingenious, and very sound. But unfortunately there is another little rule about microscopes, as Bohr pointed out. The wavelength isn't the only thing. The diameter of the objective lens is important too. For good resolution the diameter must be large. And if the diameter is large, how can we tell the direction in which the photon bounced off the electron? It might have gone through any part of the large lens we have to use to get the needed resolving power. Lenses are funny things. They bring all rays from the electron to the same focus. We cannot tell the direction of the ray by looking at the image of the electron. The principle of perversity is at work again. The larger the lens the better the resolving power, true; but also the greater the uncertainty as to the direction of the photon and thus as to the jolt it gave the electron. We cannot find the data needed for the suggested billiard ball calculation after all. The situation is as bad as ever. When we observe the position we ruin our chances of finding the velocity.

There's still a chance. What's sauce for the goose is sauce for the gander. We brought in the photons under protest. Now let us make them help us out of the hole they've put us in. Why can't we find the direction of the photon by measuring the jolt it gave the microscope?

How do we measure the jolt on the microscope? We must observe how the microscope moves. How can we observe that? By looking at it. That means shining light on it. That means bombarding it with photons. And they have to be photons of enormous energy because we are trying to measure an infinitesimal jump. This is where we came in. It is the same trouble all over again. Each time we try to bolster a previous observation with another, we cause a new jolt which makes the new information out of date. The principle of perversity is rampant and exultant. The more we strive to determine the electron's position the more we spoil our possible knowledge of its velocity; and all because of Planck's quantum h. This, more or less, is Heisenberg's justly celebrated *principle of indeterminacy*. According to this principle, we must simply reconcile ourselves to the fact that we cannot determine both the position and the velocity of a particle with exactitude, even in imagination. Now the quantum is here, we cannot know both q and p simultaneously. When we measure q we disturb p. It can be shown by other hypothetical experiments that when we measure p we disturb q. And the whole trouble is that there is no way of determining the precise amount of the disturbance. If we are content to know the position approximately rather than precisely, we may also know the momentum to some extent. It is only when we insist on knowing either the position or the momentum exactly that all vestige of information about the other is destroyed. Heisenberg found that the uncertainty in the position and the uncertainty in the momentum, if multiplied together, at the best could not give a value less than h; there is Planck's constant again, the villain of the piece.

Contrast this with the prequantum situation. There too one had to use light of small wavelength, and there too light exerted pressure. But the intensity of the light could be made as weak as one wished and the pressure thus reduced without limit; and with it the disturbance of the electron. In the quantum view, reducing the intensity does not reduce the individual jolts of the photons, it merely makes them less frequent. No observation can be made until a photon actually bounces off the electron. Since reducing the intensity does not reduce the energy of an individual photon, the jolt cannot be ignored. And the jolt itself remains essentially indeterminate.

Here is one of the significant characteristics of the new physics. There is more, and worse, to come. We have not even explained yet how this is linked to the fact that p times q and q times p are different. But let us examine for a moment what we have already found.

First we must realize that all this pictorial discussion of imaginary experiments is little more than general, even loose talk designed to make us feel more comfortable about the meaning of the quantum theory. There are no microscopes so incredibly powerful as those we have imagined. We have not even been entirely consistent, for though we regarded the electron as a particle we ended by showing that if it was a particle it was certainly a queer one. The true justification of such imaginings is the success of the quantum theory itself, for these mental experiments are but an interpretation of its basic rules.

We have come to a new concept of the particle. Whatever a particle may be, it is no longer what we used to think it was. The old particle could have position and velocity both. The new one can have position, or it can have velocity, or it can have a rather fuzzy position together with a rather fuzzy velocity, but it cannot have both together with precision. In our imaginary experiments, we thought of the electron as an old-fashioned particle, only to discover that we could not observe all its alleged attributes. But now we must renounce the old idea. The spirit of perversity will baffle us so long as we try to retain it. If the old attributes may not be observed, even in theory, we conclude that they do not really exist. We begin to envision a new type of "particle" very different from the classical idea. It cannot be regarded as a minute lump moving in a definite way. It can be regarded as a minute lump, or else as moving in a definite way, but not as both at once. Naturally, the wave-particle puzzle now takes on new significance. We shall discuss it further later.

Meanwhile there is another aspect of the present situation to be considered, for science has suddenly become more humble. In the good old days it could boldly predict the future. But what of now? To predict the future we must know the present, and the present is not knowable, for in trying to know it we inevitably alter it. If we know the airfield from which an airplane starts, and also its speed and direction, we can predict where it will be in the immediate future. But if we can know only the particular airfield, or else only the direction and speed of the plane, but not both together, then prediction becomes mere guesswork. That was the situation with the electron. Science had suffered a drastic and fundamental change without at first perceiving it. It went all the way from Planck to Heisenberg before realizing fully what had occurred; before realizing that the whole structure of scientific thought had been transformed. Its proudest boast, its most cherished illusion had been taken away from it. It had suddenly grown old and wise. It had at last realized it never had possessed the ability to predict the detailed future.

Yet science still predicts the future; and with more success than ever, thanks to the quantum. We shall clearly have to return to this matter too.

Still another item. What of the Bohr orbits? If we placed a Bohr atom under our hypothetical microscope, what would we observe? Would we be able to follow an electron around an orbit? Or even around a part of an orbit? Not at all. The very act of observation would give the electron such a jolt as would knock it from its orbit into some other permitted orbit; and in some instances would even eject it from the atom altogether. Thus, even if we permit ourselves to talk as though orbits really existed, we see that they are certainly not theoretically observable in the old sense. How sure had been Heisenberg's instinct, which led him to reject the orbits right from the start!

One final item and we may continue with our story. Just as momentum and position are paired, so too are energy and time, as Hamilton well knew. But notice now what a tremendous thing it was when Planck linked energy to frequency. We cannot measure frequency in an instant. We have to wait a little while, to watch an oscillation or two, at the least. Thus if energy is akin to frequency, we may not measure energy in an instant but must spend a little time in doing so. Compare this with what Heisenberg discovered about momentum and position and we have a perfect parallel. If we know the momentum we cannot know the exact location in space, if we know the energy we cannot know the exact location in time.

The parallel is indeed perfect. But there is a special interest for us in this relationship of time and energy, for no hypothetical microscope is needed to discover it. See how obvious it had been all along had we but the wit and courage to recognize it. There it was, crying out for recognition, as soon as Planck gave birth to the quantum; a discovery of tremendous proportions simply begging to be discovered. Anyone might have walked off with incredible scientific glory by merely pointing it out vigorously—except that no one would have taken him seriously before 1925 or thereabouts, a momentous quarter century late. Who knows but that there are similar things today, just as obvious, staring us in the face, their message disregarded because men lack the requisite daring and gallantry. For daring and gallantry are needed in science as in battle.

Let us return to our story, where much still awaits clarification. Can we explain the p times q business physically? How is it related to Heisenberg's microscope?

For this we must go back to the sign-language rules of quantum mechanics, for those rules actually implied Heisenberg's discovery. It has not escaped the reader that, for all the talk in the previous chapter, the rules of Dirac were left without formal statement. The time was not then ripe. But with Heisenberg having prepared the way for us, we may now inspect a few of their details, enough to see how admirably they encompass Heisenberg's revolutionary discovery.

According to Dirac, an electron, or atom, or nebula, or automobile, or any other dynamical system may have various possible states of motion. He represented each particular state by the symbol ψ, which, as we mentioned, was not the same as the ψ of Schrödinger. Let us fix our attention on the electron Heisenberg was subjecting to such pummeling. To observe its position we perform a certain experimental operation on it. Somehow we have to state this fact in mathematical sign language, so we denote the physical operation by a mathematical operator q. Since, as we now know, the physical operation usually disturbs the physical motion of the system, we make the mathematical operator mirror its effect by altering the ψ on which it operates; we say that q times ψ is usually different from ψ. What could be more direct than that? It is an exact replica of the corresponding physical situation in mathematical sign language.

In everyday life there are many instances of operations having different effects when performed in different orders. For instance, if we denote the operation of eating one's cake by p and that of having one's cake by q, then q may be followed by p, but not vice versa. If we denote by p the operation of washing one's hair, and by q the operation of doing something with it, then again, as every woman knows, p may follow q but q may not so easily follow p. If p denotes the operation (in either sense) of having a baby, and q that of getting married, then q followed by p is considered different from p followed by q. In all such cases we would say that p times q and q times p were different. It is much the same in quantum mechanics.

Suppose we could find the exact position of Heisenberg's electron, say that for which q has the value 3. Then we could replace the operator q by the number 3 and say that q times ψ was equal to 3 times ψ. Similarly, if we knew that the momentum had the value 5 we could say that p times ψ was equal to 5 times ψ. This is all very primitive. A mathematical savage could understand it. But such things are rules of the new quantum mechanics.

What does it mean, for Heisenberg's electron, that p times q is not the same as q times p? We are at last in a position to answer. It is a simple exercise in mathematical sign language, somewhat simpler than simple arithmetic; which only goes to show what tremendous things are hidden beneath these innocent-looking rules.

Let us for the moment pretend we found q had the definite value 3 and p the definite value 5, irrespective of which measurement was made first. Then p times q times ψ would have to be 5 times 3 times ψ, or 15 times ψ, while q times p times ψ would be 3 times 5 times ψ, or also 15 times ψ—the same as before. *And this is obviously a contradiction.* The result cannot be the same as before, for the former is p times q times ψ and the latter is q times p times ψ, and we know that p times q is not the same as q times p. To put it in a nutshell, p and q cannot both have exact numerical values, such as 5 and 3, simply because 3 times 5 and 5 times 3 are equal while such is not the case for p and q. The contradiction means, of course, that the initial assumption was false. Thus the inequality of p times q and q times p means that the order in which the observations are made affects their results. This implies that the observation of one disturbs the observation of the other, which is what Heisenberg and Bohr demonstrated with their hypothetical microscope.

Now the fun really begins. All this was but an introduction to the main revolution in scientific thought brought about by the new quantum mechanics. There is more to Heisenberg's discovery than the mere inability to know position and velocity simultaneously, and more too to the rules of quantum mechanics, correspondingly. Credulity will at first be strained to the breaking point. But there is no way out. The evidence is overwhelming. And after a while one becomes reconciled to the new ideas, however bizarre, and recognizes their probable legitimacy.

But let us go into this thing with our eyes open. Let us taste beforehand the flavor of what is to come. The situation is much as if a child had long been asking us an age-old question, seeking to learn the truth.

"Daddy," she says, "which came first, the chicken or the egg?"

Steadfastly, even desperately, we have been refusing to commit ourselves. But our questioner is insistent. The truth alone will satisfy her. Nothing less. At long last we gather up courage and issue our solemn pronouncement on the subject:

"Yes!"

So it is here.

"Daddy, is it a wave or a particle?"

"Yes."

"Daddy, is the electron here or is it there?"

"Yes."

"Daddy, do scientists really know what they are talking about?"

"Yes!"

The way has already been prepared. We already know from Heisenberg's principle that a particle is no longer what it used to be. We are about to find out that it is even less like its old self than we think even now. Because the word "particle" is now ambiguous, and contaminated by its classical associations, we shall talk rather of an electron though what we say is applicable to a photon, an atom, or any other "particle." The word "electron," however, is itself somewhat contaminated. We usually think of it as a particle of the older sort, and we must realize that this is a major reason why everything will seem strange and paradoxical. Despite all paradox, however, we must always keep before us the realization that we are talking of the world, and not of idle theories spun out of gossamer. We are talking of what you and I are made of, and the trees and the stones, the stars and the atomic bombs, radio waves and viruses, and cabbages and kings; and, for all we know, we are talking of the material basis of love and hate, and patriotism and treachery, and religious ecstasy. Behind all our quaint ideas about p's and q's and their indeterminacies lies a world of harsh reality in proved relation to them.

Bearing all this in mind, how would we feel to be told that an electron can be in two places at once, or can be going in two directions at the same time; and more than two? We shall soon learn, for something closely akin to this must now be swallowed, just as it had to be by physicists not so many years ago. It can all be made into a consistent scheme in the end, however, and on the way can be related to analogous things already familiar to us.

First, then, let us get the worst over. After that we may see how to make the best of it, and in so doing reach a position so comfortable that we would hesitate to retreat from it to the older point of view with its inability to account for some of the outstanding basic experiments.

An electron might be moving straight upwards at five thousand feet per second. That would be a legitimate state of motion and could be denoted by ψ. An electron might be moving to the right at eight hundred feet per second. That also would be a state of motion, representable by a different ψ. Now we must remember that Dirac's sign-language rules were distilled from the successful theories of Heisenberg and Schrödinger. One of these rules, perhaps the most important, is the rule of superposition. We have not mentioned it before. It tells us we could have a state of motion consisting of a combination of the two states above, so much of the first and so much of the second. This is something utterly radical. It does not mean a classical motion intermediate between the two, as simultaneous motions north and east combine classically to form a single northeasterly motion. It does not mean anything so convenient and snug as that. It means *both motions at once.*

For the time being we can console ourselves with the thought that we are dealing with probabilities, but this consolation, though correct, will not last long without modification. Let us try it, even so, for it is an important part of the new physics.

Suppose we start with an electron moving upward at five thousand feet per second. If we observe its position we operate on its ψ with q. The mathematics then shows that the new state, q times ψ, is a combination state consisting not of just two but of an infinite number of pure motions all going on at once. To see what this may mean physically, let us look through Heisenberg's microscope. In observing the position of the electron we have made an indeterminate change in its velocity. All we can now know about its motion, therefore, is that it is probably such and such a motion, or with less likelihood another motion, or another, or another, through an infinite list of possibilities. Though we can make a catalogue of all the possible motions and even determine their relative likelihoods, we cannot fix the exact motion without making a further observation—and that would not help, because it alters the motion and renders out of date the information it itself yields. In this sense, then, the electron is in several states of motion at once; in the sense that it is really in one particular one, but that we do not and cannot know which. Remember, things are not going to remain quite so simple as this. But this is a good stepping-stone to deeper waters. Let us rest on it a while and look around for familiar landmarks, and everyday analogies.

We find a somewhat similar situation in heredity. To take a simple instance, let a black fowl and a white fowl mate and produce a chicken. Before the egg is hatched we do not know its color. But it is possible to say that the chicken is in some sort of combination state of color, being twenty-five per cent black, and twenty-five per cent white, and fifty per cent that gray, bluish type known as Andalusian.

What does this mean? Not that the chicken is somehow all these at once. It is only one of them. But, lacking the complete information, we must content ourselves with the probabilities. If a hundred chickens came from such parentage, we would expect, from past experience, that about twenty-five of them would be black, twenty-five white, and fifty Andalusian. A particular chicken is one particular color. When it emerges from its egg we can determine which color it is. This, however, increases our information. It corresponds to an observation, and alters the state from a combination state to one pertaining to a particular color.

This idea that a combination state represents a lack of information may be illustrated by the hunting of a submarine without benefit of radar or other instruments of detection. An aviator observing a submarine just as it completes the act of submerging, knows its position, but not how it is moving. Therefore, for him, the submarine is in a combination state comprising all possible motions away from the point of submersion.

We can push this last analogy further. The submarine, in submerging, has merged itself with the waves in a double sense. It has become a Schrödinger wave packet.

When Schrödinger formed his electron waves into a wave packet he managed to give the electron location. But the wave packet would not stay together. It spread out and flattened. Why? Because, according to Heisenberg's principle, as soon as Schrödinger gave his electron location he lost information as to its motion. This all ties in with Born's idea that the Schrödinger waves are waves of probability. For where was the electron a few moments after it was located if its motion was unknown? In a sense, it was nowhere in particular. It might be almost anywhere, though most likely somewhere in the vicinity of the original position. Its position was now little more than a rapidly spreading probability. Its probability wave packet was spreading. As time went on, the ignorance of its position increased. Its wave packet spread further.

It is the same with the submarine. When it is observed in the act of submerging its wave packet is at its peak. If the plane reaches the spot quickly it can drop depth charges with a good chance of success. But should the plane delay, its chances of scoring a hit become less and less, for as time goes on the "position" of the submarine spreads out as an ever-widening circular region of probability, just as the ripples on the surface of the ocean spread out from its point of submersion. The submarine's wave packet, originally at its peak, spreads rapidly apart as precious time speeds on.

We shall return to the wave packets. We have not done with them yet. But there are other aspects and analogies to consider; such as the tossing of a coin. When we flip a coin, it is neither heads nor tails until it actually lands. While it is in the air it is twirling rapidly. But now suppose we could have no knowledge at all of what occurs between the tossing and the final landing. Suppose the world were so constituted that no observation of the intermediate motion was possible. Suppose some pigheaded principle of perversity prevented any such observation. What sort of theory would we form of the flipping of a coin?

Surely one thing would strike us at once as of outstanding importance: the coin could be only heads or tails, and nothing else. There were only two possible results to an observation. And we would soon find there was no way of telling beforehand which result would turn up. If we decided to confess our ignorance on this score in the language of the new physics we would say that the state of the coin was a combination of heads and tails. Since a long series of observations would show that the coins came down heads about as frequently as they came down tails, we would say that in the combination state the coin had a fifty per cent probability of being in the heads position and a fifty per cent probability of being in the tails position. As soon as the coin came to rest on the table, of course, we would know quite definitely which it was: either heads or tails. The state would be changed from a combination to a pure state, and it would be changed by the very act of observation; all of which we could express in mathematical sign language.

Suppose we were of a visual turn of mind. Then we would try to imagine what was going on in pictorial terms. We would try to imagine some intermediate mechanism or process. If we were really clever we might even imagine the coin could be twirling. That would be a satisfactory picture, and certainly would not contradict any of the known effects which the principle of perversity permitted our observing. The only trouble would be that we had no way of observing the actual twirling itself. But we could bolster our confidence with a little quantum theory. For we could note that energy was imparted to the coin to make it twirl. Energy is allied to frequency, and the frequency could well be the rate at which the coin was turning, the greater the energy the higher being the rate of turning.

With this picture in mind we would suddenly have a flash of inspiration. Why did we never find actual evidence of the twirling? Obviously, because the only mode of observation permitted us was to let the coin come to rest on the table. The act of observation thus gave the coin an undetermined jolt. The motion of the coin was perfectly free; we ourselves were responsible for making it appear to have only two possible positions. It was letting the coin fall on the table that forced it to be either heads or tails and nothing in between. If we had let it continue falling it would have continued twirling— though, of course, the principle of perversity would then not let us observe it at all.

Soon, however, we would come to realize that all this was only a mental picture designed to make us feel comfortable. It left out the one thing above all others which must be retained. It left out the very principle of perversity itself. The twirling was not observable. For all we know, it did not really take place. If it did take place, the principle of perversity effectively prevented our seeing it. If the principle of perversity was anything more than a coincidence, and its malign persistence would certainly indicate it was something far more potent and fundamental, then we must be wary of introducing the twirling motion it persisted in hiding from us, for perhaps there was no such motion after all. Though we could explain why the jolt of observation would always mask the twirling, that did not mean the twirling actually existed. To argue like that would be like asserting there was maybe a lovely design in red and green on the coin, but unfortunately it so happened we were red-green color blind. Until some experimental way around the principle of perversity was discovered, the twirling would not be a reliable object of scientific thought. We must play safe or we might be misled. We already have a perfectly adequate theory which covers all the observed facts. Why should we wish to go further? We must return to the austere point of view, and not attempt to picture a twirling motion or any other such intermediate mechanism. We must go back to our idea of the coin being in two states of position at once, part heads and part tails, and of the state being changed by the observation.

Naturally we part with our mental picture of a twirling coin with considerable reluctance and regret. It was such fun while it lasted. Perhaps we retain a lingering hope that it is not gone forever. Who knows but that some great scientific advance may some day make the twirling visible for all to see? But until that hypothetical day, all this is whimsey—perhaps even dangerous whimsey. For the joke is that we really do not know whether the coin was twirling at all. If perversity prevented our observing it, why could it not also prevent the twirling itself? Our sign-language theory would still apply, twirling or no twirling, for it was based only on known results. Our picture of a twirling coin, however, is obviously a pure conjecture.

Do we still wish to cling to the twirling? Do we think there is no other possible explanation that would make sense? Does it seem that we have been splitting philosophical hairs to pretend the twirling might be illusory? Then let us think of a commonplace occurrence with which we can hardly fail to be familiar. When we get a busy signal from a pay telephone and our nickel is returned, do we really think there was a twirling all the time the coin was in the box? We could make quite an ingenious theory connecting the twirling with the ringing sound we heard through the receiver. It would account nicely for the fact that the ringing ceased when the coin reappeared. It would be wrong, nevertheless. Why, for all we know, it was not even our own nickel that was returned. It would not be hard to imagine within the phone a reservoir of nickels from which one was dropped when the receiver was replaced. We have to be careful about jumping to conclusions. Though they may seem perfectly obvious they may none the less be wrong. The twirling was pure conjecture after all.

We must look on Heisenberg's principle of indeterminacy in this light. For all that that principle makes the particle seem a genuine old-fashioned particle by placing the full blame for its idiosyncrasies upon the unavoidable clumsiness of the experimenter, it does not validate the old idea of a particle. On the contrary, the fact that the clumsiness is unavoidable and indeterminate points up the ubiquity and power of that spirit of perversity which dogs our attempts to observe the full attributes of the particle, if old-fashioned particle it really be, and casts grave doubt on its old-fashioned pretensions. Much mental confusion can arise from not heeding this.

The time has now come to leave our comfortable stepping stone. The final straw must now be gently added. When we say we have an electron in a combination state, going both north and east simultaneously, we would like this to be a simple confession of ignorance. We would like it to mean that the electron is really going due north, or else due east, but all we know for certain is that there is such and such a probability of its doing the former and such and such a probability of its doing the latter. We know these probabilities from having performed the identical experiment many times before. But no matter how carefully we work, the experiments do not yield an unequivocal answer; only the two probabilities. A single electron, we would like to say, is performing only one of these two possible motions, there being, however, no way of telling which one without performing another, different experiment and thus changing the state.

Alas, it will not do. We have to spoil it all. We cannot

maintain our convenient fiction against the pressure of experimental facts, for there is still the wave-particle battle to be resolved.

Let us look once more at the basic armaments of the wave and particle. Do we wish to prove the electron a particle? All right. We let it strike a fluorescent screen and observe its tiny scintillation; or we watch its slender track in a cloud chamber; or we let it fall on a photographic plate and note the small spot that appears on development. Behold, we have a particle.

Do we wish to prove the electron a wave? All right. We set up a screen with two pinholes in it close together, we let electrons stream through them from a single source, and we point with pride, not unmixed with smug self-satisfaction, to the characteristic interference pattern on the photographic plate beyond the screen. Behold, we have a wave.

To add to the interest, let us combine two demonstrations so that at the same time we prove the electron a wave we also prove it a particle. That will give us something really worth thinking about. All we need to do is send electrons from a single source through two pinholes in a screen and allow them to fall on a scintillation screen on the other side. Then the scintillations show we have particles while the interference patterns show we have waves; a quite fantastic situation.

But something begins to excite our suspicions. The waves seem to come from crowds of electrons rather than from individual ones. Don't let the crowds of electrons confuse us. Let us watch carefully what one single electron does. If it ultimately produces a scintillation it is surely a particle. How can it then also be a wave? Because it produced an interference pattern? What interference pattern? One solitary scintillation is not an interference pattern. The interference pattern is produced by a vast crowd of scintillations. It is an effect pertaining to the multitude. The individual scintillations pile up in some places and not in others, that is all. When artillery lays down a barrage pattern, the pattern is not discernible in a single shell burst, but only in a group of them. We could easily lay down a barrage that would give the appearance of an interference pattern; yet it would not mean the shell was a wave. When breezes blow over a field of wheat we have all seen the speeding waves course over its surface. Yet the wheat is not a wave. At last we have solved the problem of wave and particle. The electron is after all a particle, and so is the photon. It looks like a wave only when observed in enormous crowds.

But this will not do. We are only misleading ourselves. We are still trying to escape the inexorable conclusion. Already we are heading in the wrong direction, and we are in danger of making the fatal blunder of underestimating the subtlety of our problem. It is not of so naïve a sort as this, else it would have been solved long before. True, the interference pattern is manifest when we have a crowd of electrons. But there must be some cause of the interference pattern even so. And this cause must lie within each single electron. The pattern is not just a crowd effect. The crowd is merely what makes it easy to see. Somehow the pattern is latent in each individual electron. If we fire our artillery shells one after the other, instead of many at a time, we can still produce the same pattern of shell craters as before. That is because a human agency directs the firing. If we send our electrons out less and less frequently and take note of their individual scintillations, falling one at a time, we also find them still conforming to the proper interference pattern, but this is much harder to understand than the artillery pattern, for there is no obvious external agency directing the electron gunfire according to a preconceived pattern. Though each individual scintillation seems to fall at random, there is a subtle architecture in this randomness, for the scintillations gradually build up to the characteristic pattern of interference.

How does the electron do it? Which hole in the screen did any particular electron go through? The interference pattern is a two-hole pattern, quite different from a one-hole pattern. There is no possible escape. The grim conclusion is unavoidable. Whether we like it or not, if a single electron somehow contains within itself the two-hole interference pattern, *that single electron must have passed through both holes; and after passing through both holes it must have interfered with itself!* That is the revolutionary and well-nigh intolerable conclusion which experiment forces upon us.

Is it too much to swallow? Is it incredible? Is it against common sense? Perhaps. Yet it is based on the strongest scientific evidence.

Wait! We will fool it. We will make it confess its own falsity. We will place a recording device at each hole in the screen. Then if we send out one single electron from our source and it passes through the screen we must surely detect it going through one or the other hole and not through both, for we know we never observe a fraction of an electron. That way we will prove definitely it went through only one of the holes, and will even name which one it went through. We are not ones to be so easily fooled with impossible theories. We are not children, believing in fairy tales. Enough of all such nonsense.

Yes, it is true that we can discover in this way which hole

the electron went through, and can even show that it went through one hole only and not both. But that would be an entirely different experiment. It would not contradict what we said above, for we would no longer be passing electrons through a screen with two simple holes in it. The spirit of perversity is always on the job. It never sleeps. Let us watch it at its fascinating work here. Suppose we find that the electron went through the lower hole. Since the recording instrument at that hole was affected by the electron, the electron must have been affected by the instrument, the precise effect on the electron being indeterminate. What hope is there of obtaining an interference pattern with a crowd of electrons if each electron is affected differently and arbitrarily as it goes through the screen? If we cannot now produce a two-hole interference pattern, what need is there now to claim that each electron went through both holes? The whole situation is vastly different from before. In closing one door of our trap we have had to open another. The very device that shows that no single electron went through both holes at once itself destroys the two-hole interference pattern, thereby letting the electron escape the trap.

Or look at this from another point of view. When both holes are unencumbered, any electrons which traverse the screen must be in a combination state of motion, going through both holes at once. The two motions interfere with each other to produce the interference patterns. What happens when we introduce our recording devices? Any electron which now traverses the screen is fixed in a pure state of motion, passing either through one hole or else through the other. We may no longer expect a two-hole interference pattern, for we have made an additional observation and thus altered the state of motion so that it no longer pertains to two holes.

Is all this difficult and discouraging? Does the idea of an electron in several places at once or with several states of motion at once give us pause? Does it revolt our sensibilities? We have been too particular. We have leaned too heavily on the particle image. Let us not imagine that scientists accepted these new ideas with cries of joy. They fought them and resisted them as much as they could, inventing all sorts of traps and alternative hypotheses in vain attempts to escape them. But the glaring paradoxes were there as early as 1905 in the case of light, and even earlier, and no one had the courage or wit to resolve them until the advent of the new quantum mechanics. The new ideas are so difficult to accept because we still instinctively strive to picture them in terms of the old-fashioned particle, despite Heisenberg's indeterminacy principle. We still shrink from visualizing an electron as something which having motion may have no position, and having position may have no such thing as motion or rest. We still try to blame the clumsiness of the innocent experimenter for this fundamental characteristic of the electron, or the photon.

We have not abandoned our former steppingstone in all this, but rather have made it a base for further advance. We may still look on a combination state of motion as a confession of our ignorance as to the precise outcome of an observation, and may still regard it as listing various probabilities. The interference patterns, embodiments of these probabilities, are still discernible only as crowd effects. It is the mental picture that has changed. We have learned at last the sheer impossibility of visualizing atomic processes except in terms of the most grotesque images. We have seen what fantastic shapes our mental images must take if they would spy on that which the principle of indeterminacy veils.

It was Bohr who realized these things most surely and profoundly. He it was who finally resolved the wave-particle conflict, and first delineated with fundamental clarity an outline of the puzzling new era in science. He it was who saw that the wave and particle were but two aspects of the same thing. They were not enemies. Their whole battle had been a sham. Their persistent warfare had been one long fraud, a superb example of the power of classical propaganda. If the wave collared a piece of territory, the particle never really disputed it, but opened up a new region of its own. If the wave explained interference, the particle took no serious counteraction but consoled itself with staking a claim to the photoelectric effect, a claim never contested by the wave. It had been the most polite type of pseudo warfare imaginable, but done up with such bellicose classical trumpetings as to give the false impression of terrible battle. What happened, for example, when we placed indicating devices at the holes in the screen? Did they force the wave and particle into genuine battle? Not at all. The particle politely found a way for the wave to escape the trap without embarrassment.

When scientists at last suspected the true nature of these antics they devised sterner, more devilish tricks to make the wave and particle join battle. But Bohr and others were able to prove in detail that the gentle spirit of perversity, Heisenberg's principle of indeterminacy, was ever alert to prevent even the beginning of strife. If we try to regard the wave and particle as two distinct entities, we must think of them not as implacable feudists but as professional wrestlers putting on a show. But they are really not distinct. They are alternative, partial images of the selfsame thing.

This complementary aspect of particle and wave is a central

feature of the new physics. It is inescapable; part of the very fabric of quantum mechanics. The sign-language rules require it, and Heisenberg's indeterminacy principle offers a pictorial justification. Some pages back we pretended that the Schrö-dinger wave packets were no more than a superficial analogy of the mixed states of Dirac. It was Bohr, primarily, who revealed that they are far more than an analogy. They are, in fact, an exact counterpart, but expressed in the language of waves rather than in that of particles. From them one may readily extract the indeterminacy relation of momentum and position, or of energy and time. Indeed, we have already indi-cated as much, for, having discussed the indeterminacy of momentum and position from the particle aspect, did we not play fair by inferring the indeterminacy of energy and time from the point of view of the wave? Just as the theories of Schrödinger and Heisenberg merge into the single theory of Dirac, so do the wave and particle merge into a single self-consistent whole; an entity for which Eddington aptly proposed the name "wavicle."

> There was a little girl and she had a little curl
> Right in the middle of her forehead.
> When she was good, she was very very good
> But when she was bad she was horrid.

Like the little girl with the curl, the electron sometimes shows one side of its nature and sometimes the other. It is still an electron for all that, and a perfectly normal and healthy one. It would not be an electron did it not display a well-rounded personality, being sometimes like a wave and sometimes like a particle. If red light shines on the pages of this book, the paper appears bright red; but if we change over to blue light, the bright red changes to blue. There is no contradiction here. The early redness of the paper does not contradict its later blueness, any more than the varied colors of the sunset contradict the brilliance of high noon. It is our manner of observation that has changed. And the very change from red light to blue prevents our continuing to observe the paper as red. Whether we find the electron in its wave or its particle aspect depends similarly, and without contradiction, on the way we observe it. Just as we can make the little girl with the curl very very good by letting her show off and scintillate, or make her bad and horrid by interference, a change so great we would hardly recognize her as the same child, so too can we put the electron into a particle mood by letting it scintillate or into a wave mood by seeking interference. Through it all, whether wave or particle, it remains an electron. Like the photon, it remains a wavicle.

"Daddy, is the electron here or is it there?"

"Yes."

"Daddy, is it a wave or a particle?"

"Yes."

How honest we were with our little questioner, after all! To see if she has learned the lesson we may ask her a question in return:

"Is a mermaid a woman or a fish?"

She should have no difficulty deciding on the appropriate answer.

CHAPTER XIV

THE NEW LANDSCAPE OF SCIENCE

LET US now gather the loose threads of our thoughts and see what pattern they form when knit together.

We seem to glimpse an eerie shadow world lying beneath our world of space and time; a weird and cryptic world which somehow rules us. Its laws seem mathematically precise, and its events appear to unfold with strict causality.

To pry into the secrets of this world we make experiments. But experiments are a clumsy instrument, afflicted with a fatal indeterminacy which destroys causality. And because our mental images are formed thus clumsily, we may not hope to fashion mental pictures in space and time of what transpires within this deeper world. Abstract mathematics alone may try to paint its likeness.

With indeterminacy corrupting experiment and dissolving causality, all seems lost. We must wonder how there can be a rational science. We must wonder how there can be anything at all but chaos. But though the detailed workings of the indeterminacy lie hidden from us, we find therein an astounding uniformity. Despite the inescapable indeterminacy of experiment, we find a definite, authentic residue of exactitude and determinacy. Compared with the detailed determinacy claimed by classical science, it is a meager residue indeed. But it is precious exactitude none the less, on which to build a science of natural law.

The very nature of the exactitude seems a paradox, for it is an exactitude of probabilities; an exactitude, indeed, of wave-like, interfering probabilities. But probabilities are potent things—if only they are applied to large numbers. Let us see what strong reliance may be placed upon them.

When we toss a coin, the result may not be predicted, for it is a matter of chance. Yet it is not entirely undetermined. We know it must be one of only two possibilities. And, more important even than that, if we toss ten thousand coins we know we may safely predict that about half will come down heads. Of course we might be wrong once in a very long while. Of course we are taking a small risk in making such a prediction. But let us face the issue squarely, for we really place far more confidence in the certainty of probabilities than we sometimes like to admit to ourselves when thinking of them abstractly. If someone offered to pay two dollars every time a coin turned up heads provided we paid one dollar for every tails, would we really hesitate to accept his offer? If we did hesitate, it would not be because we mistrusted the probabilities. On the contrary, it would be because we trusted them so well we smelled fraud in an offer too attractive to be honest. Roulette casinos rely on probabilities for their gambling profits, trusting to chance that, in the long run, zero or double zero will come up as frequently as any other number and thus guarantee them a steady percentage of the total transactions. Now and again the luck runs against them and they go broke for the evening. But that is because chance is still capricious when only a few hundred spins are made. Insurance companies also rely on probabilities, but deal with far larger numbers. One does not hear of their ever going broke. They make a handsome living out of chance, for when precise probabilities can be found, chance, in the long run, becomes practical certainty. Even classical science built an elaborate and brilliantly successful theory of gases upon the seeming quicksands of probability.

In the new world of the atom we find both precise probabilities and enormous numbers, probabilities that follow exact mathematical laws, and vast, incredible numbers compared with which the multitude of persons carrying insurance is as nothing. Scientists have determined the weight of a single electron. Would a million electrons weigh as much as a feather, do you think? A million is not large enough. Nor even a billion. Well, surely a million billion then. No. Not even a billion billion electrons would outweigh the feather. Nor yet a million billion billion. Not till we have a billion billion billion can we talk of their weight in such everyday terms. Quantum mechanics having discovered precise and wonderful laws governing the probabilities, it is with numbers such as these that science overcomes its handicap of basic indeterminacy. It is by this means that science boldly predicts. Though now humbly confessing itself powerless to foretell the exact behavior of individual electrons, or photons, or other fundamental entities, it yet can tell with enormous confidence how such great multitudes of them must behave precisely.

But for all this mass precision, we are only human if, on first hearing of the breakdown of determinacy in fundamental science, we look back longingly to the good old classical days, when waves were waves and particles particles, when the workings of nature could be readily visualized, and the future was predictable in every individual detail, at least in theory. But the good old days were not such happy days as nostalgic, rose-tinted retrospect would make them seem. Too many contradictions flourished unresolved. Too many well-attested facts played havoc with their pretensions. Those were but days of scientific childhood. There is no going back to them as they were.

Nor may we stop with the world we have just described, if we are to round out our story faithfully. To stifle nostalgia, we pictured a world of causal law lying beneath our world of space and time. While important scientists seem to feel that such a world should exist, many others, pointing out that it is not demonstrable, regard it therefore as a bit of homely mysticism added more for the sake of comfort than of cold logic.

It is difficult to decide where science ends and mysticism begins. As soon as we begin to make even the most elementary theories we are open to the charge of indulging in metaphysics. Yet theories, however provisional, are the very lifeblood of scientific progress. We simply cannot escape metaphysics, though we can perhaps overindulge, as well as have too little. Nor is it feasible always to distinguish good metaphysics from bad, for the "bad" may lead to progress where the "good" would tend to stifle it. When Columbus made his historic voyage he believed he was on his westward way to Japan. Even when he reached land he thought it was part of Asia; nor did he live to learn otherwise. Would Columbus have embarked upon his hazardous journey had he known what was the true westward distance of Japan? Quantum mechanics itself came partly from the queer hunches of such men as Maxwell and Bohr and de Broglie. In talking of the meaning of quantum mechanics, physicists indulge in more or less mysticism according to their individual tastes. Just as different artists instinctively paint different likenesses of the same model, so do scientists allow their different personalities to color their interpretations of quantum mechanics. Our story would not be complete did we not tell of the austere conception of quantum mechanics hinted at above, and also in our parable of the coin and the principle of perversity, for it is a view held by many physicists.

These physicists are satisfied with the sign-language rules, the extraordinary precision of the probabilities, and the strange, wavelike laws which they obey. They realize the impossibility of following the detailed workings of an indeterminacy through which such bountiful precision and law so unaccountably seep. They recall such incidents as the vain attempts to build models of the ether, and their own former naïve beliefs regarding momentum and position, now so rudely shattered. And, recalling them, they are properly cautious. They point to such things as the sign-language rules, or the probabilities and the exquisite mathematical laws in multidimensional fictional space which govern them and which have so eminently proved themselves in the acid test of experiment. And they say that these are all we may hope and reasonably expect to know; that science, which deals with experiments, should not probe too deeply beneath those experiments for such things as cannot be demonstrated even in theory.

The great mathematician John von Neumann, who accomplished the Herculean labor of cleaning up the mathematical foundations of the quantum theory, has even proved mathematically that the quantum theory is a complete system in itself, needing no secret aid from a deeper, hidden world, and offering no evidence whatsoever that such a world exists. Let us then be content to accept the world as it presents itself to us through our experiments, however strange it may seem. This and this alone is the image of the world of science. After castigating the classical theorists for their unwarranted assumptions, however seemingly innocent, would it not be foolish and foolhardy to invent that hidden world of exact causality of which we once thought so fondly, a world which by its very nature must lie beyond the reach of our experiments? Or, indeed, to invent anything else which cannot be demonstrated, such as the detailed occurrences under the Heisenberg microscope and all other pieces of comforting imagery wherein we picture a wavicle as an old-fashioned particle preliminary to proving it not one?

All that talk of exactitude somehow seeping through the indeterminacy was only so much talk. We must cleanse our minds of previous pictorial notions and start afresh, taking the laws of quantum mechanics themselves as the basis and the complete outline of modern physics, the full delineation of the quantum world beyond which there is nothing that may properly belong to physical science. As for the idea of strict causality, not only does science, after all these years, suddenly find it an unnecessary concept, it even demonstrates that according to the quantum theory strict causality is fundamentally and intrinsically undemonstrable. Therefore, strict causality is no longer a legitimate scientific concept, and must be cast out from the official domain of present-day science. As Dirac has written, "*The only object of theoretical physics is to calculate results that can be compared with experiment*, and it is quite unnecessary that any satisfying description of the whole course of the phenomena should be given." The italics here are his. One cannot escape the feeling that it might have been more appropriate to italicize the second part of the statement rather than the first!

Here, then, is a more restricted pattern which, paradoxically, is at once a more cautious and a bolder view of the world of quantum physics; cautious in not venturing beyond what is well established, and bold in accepting and being well content with the result. Because it does not indulge too freely in speculation it is a proper view of present-day quantum physics, and it seems to be the sort of view held by the greatest number. Yet, as we said, there are many shades of opinion, and it is sometimes difficult to decide what are the precise views of particular individuals.

Some men feel that all this is a transitional stage through which science will ultimately pass to better things—and they hope soon. Others, accepting it with a certain discomfort, have tried to temper its awkwardness by such devices as the introduction of new types of logic. Some have suggested that the observer creates the result of his observation by the act of observation, somewhat as in the parable of the tossed coin. Many nonscientists, but few scientists, have seen in the new ideas the embodiment of free will in the inanimate world, and have rejoiced. Some, more cautious, have seen merely a revived possibility of free will in ourselves now that our physical processes are freed from the shackles of strict causality. One could continue endlessly the list of these speculations, all testifying to the devastating potency of Planck's quantum of action h, a quantity so incredibly minute as to seem utterly inconsequential to the uninitiated.

That some prefer to swallow their quantum mechanics plain while others gag unless it be strongly seasoned with imagery and metaphysics is a matter of individual taste behind which lie certain fundamental facts which may not be disputed; hard, uncompromising, and at present inescapable facts of experiment and bitter experience, agreed upon by all and directly opposed to the classical way of thinking:

There is simply no satisfactory way at all of picturing the fundamental atomic processes of nature in terms of space and time and causality.

The result of an experiment on an individual atomic particle generally cannot be predicted. Only a list of various possible results may be known beforehand.

Nevertheless, the statistical result of performing the same individual experiment over and over again an enormous number of times may be predicted with virtual certainty.

For example, though we can show there is absolutely no contradiction involved, we cannot visualize how an electron which is enough of a wave to pass through two holes in a screen and interfere with itself can suddenly become enough of a particle to produce a single scintillation. Neither can we predict where it will scintillate, though we can say it may do so only in certain regions but not in others. Nevertheless when, instead of a single electron, we send through a rich and abundant stream we can predict with detailed precision the intricate interference pattern that will build up, even to the relative brightness of its various parts.

Our inability to predict the individual result, an inability which, despite the evidence, the classical view was unable to tolerate, is not only a fundamental but actually a plausible characteristic of quantum mechanics. So long as quantum mechanics is accepted as wholly valid, so long must we accept this inability as intrinsically unavoidable. Should a way ever be found to overcome this inability, that event would mark the end of the reign of quantum mechanics as a fundamental pattern of nature. A new, and deeper, theory would have to be found to replace it, and quantum mechanics would have to be retired, to become a theory emeritus with the revered, if faintly irreverent title "classical."

Now that we are accustomed, a little, to the bizarre new ideas we may at last look briefly into the quantum mechanical significance of something which at first sight seems trivial and inconsequential, namely, that electrons are so similar we cannot tell one from another. This is true also of other atomic particles, but for simplicity let us talk about electrons, with the understanding that the discussion is not thereby confined to them alone.

Imagine, then, an electron on this page and another on the opposite page. Take a good look at them. You cannot tell them apart. Now blink your eyes and take another look at them. They are still there, one on this page and one on that. But how do you know they did not change places just at the moment your eyes were closed? You think it most unlikely? Does it not always rain on just those days when you go out and leave the windows open? Does it not always happen that your shoelace breaks on just those days when you are in a special hurry? Remember these electrons are identical twins and apt to be mischievous. Surely you know better than to argue that the electron interchange was unlikely. You certainly could not prove it one way or another.

Perhaps you are still unconvinced. Let us put it a little differently, then. Suppose the electrons collided and bounced off one another. Then you certainly could not tell which one was which after the collision.

You still think so? You think you could keep your eyes

glued on them so they could not fool you? But, my dear sir, that is classical. That is old-fashioned. We cannot keep a continual watch in the quantum world. The best we can do is keep up a bombardment of photons. And with each impact the electrons jump we know not how. For all we know they could be changing places all the time. At the moment of impact especially the danger of deception is surely enormous. Let us then agree that we can never be sure of the identity of each electron.

Now suppose we wish to write down quantum equations for the two electrons. In the present state of our theories, we are obliged to deal with them first as individuals, saying that certain mathematical co-ordinates belong to the first and certain others to the second. This is dishonest though. It goes beyond permissible information, for it allows each electron to preserve its identity, whereas electrons should belong to the nameless masses. Somehow we must remedy our initial error. Somehow we must repress the electrons and remove from them their unwarranted individuality. This reduces to a simple question of mathematical symmetries. We must so remold our equations that interchanging the electrons has no physically detectable effect on the answers they yield.

Imposing this nonindividuality is a grave mathematical restriction, strongly influencing the behavior of the electrons. Of the possible ways of imposing it, two are specially simple mathematically, and it happens that just these two are physically of interest. One of them implies a behavior which is actually observed in the case of photons, and α particles, and other atomic particles. The other method of imposing nonindividuality turns out to mean that the particles will shun one another; in fact, it gives precisely the mysterious exclusion principle of Pauli.

This is indeed a remarkable result, and an outstanding triumph for quantum mechanics. It takes on added significance when we learn that all those atomic particles which do not obey the Pauli principle are found to behave like the photons and α particles. It is about as far as anyone has gone toward an understanding of the deeper significance of the exclusion principle. Yet it remains a confession of failure, for instead of having nonindividuality from the start we begin with individuality and then deny it. The Pauli principle lies far deeper than this. It lies at the very heart of inscrutable Nature. Someday, perhaps, we shall have a more profound theory in which the exclusion principle will find its rightful place. Meanwhile we must be content with our present veiled insight.

The mathematical removal of individuality warps our equations and causes extraordinary effects which cannot be properly explained in pictorial terms. It may be interpreted as bringing into being strange forces called exchange forces, but these forces, though already appearing in other connections in quantum mechanics, have no counterpart at all in classical physics.

We might have suspected some such forces were involved. It would have been incredibly naïve to have believed that so stringent an ordinance against overcrowding as the exclusion principle could be imposed without some measure of force, however well disguised.

Is it so sure that these exchange forces cannot be properly explained in pictorial terms? After all, with force is associated energy. And with energy is associated frequency according to Planck's basic quantum law. With frequency we may associate some sort of oscillation. Perhaps, then, if we think not

of the exchange forces themselves but of the oscillations associated with them we may be able to picture the mechanism
through which these forces exist. This is a promising idea. But
if it is clarity we seek we shall be greatly disappointed in it.

It is true there is an oscillation involved here, but what
a fantastic oscillation it is: a rhythmic interchange of the electrons' identities. The electrons do not physically change places
by leaping the intervening space. That would be too simple.
Rather, there is a smooth ebb and flow of individuality between
them. For example, if we start with electron A here and electron B on the opposite page, then later on we would here have
some such mixture as sixty per cent A and forty per cent B,
with forty per cent A and sixty per cent B over there. Later still
it would be all B here and all A there, the electrons then
having definitely exchanged identities. The flow would now
reverse, and the strange oscillation continue indefinitely. It is
with such a pulsation of identity that the exchange forces of
the exclusion principle are associated. There is another type
of exchange which can affect even a single electron, the electron being analogously pictured as oscillating in this curious,
disembodied way between two different positions.

Perhaps it is easier to accept such curious pulsations if we
think of the electrons more as waves than as particles, for then
we can imagine the electron waves becoming tangled up with
each other. Mathematically this can be readily perceived, but
it does not lend itself well to visualization. If we stay with the
particle aspect of the electrons we find it hard to imagine what
a 60 per cent–40 per cent mixture of A and B would look like
if we observed it. We cannot observe it, though. The act of
observation would so jolt the electrons that we would find
either pure A or else pure B, but never a combination, the
percentages being just probabilities of finding either one. It

is really our parable of the tossed coin all over again. In mid-air the coin fluctuates rhythmically from pure heads to pure tails through all intermediate mixtures. When it lands on the table, which is to say when we observe it, there is a jolt which yields only heads or tails.

Though we can at least meet objections, exchange remains an elusive and difficult concept. It is still a strange and awe-inspiring thought that you and I are thus rhythmically exchanging particles with one another, and with the earth and the beasts of the earth, and the sun and the moon and the stars, to the uttermost galaxy.

A striking instance of the power of exchange is seen in chemical valence, for it is essentially by means of these mysterious forces that atoms cling together, their outer electrons busily shuttling identity and position back and forth to weave a bond that knits the atoms into molecules.

Such are the fascinating concepts that emerged from the quantum mechanical revolution. The days of tumult shook science to its deepest foundations. They brought a new charter to science, and perhaps even cast a new light on the significance of the scientific method itself. The physics that survived the revolution was vastly changed, and strangely so, its whole outlook drastically altered. Where once it confidently sought a clear-cut mechanical model of nature for all to behold, it now contented itself with abstract, esoteric forms which may not be clearly focused by the unmathematical eye of the imagination. Is it as strongly confident as once it seemed to be in younger days, or has internal upheaval undermined its health and robbed it of its powers? Has quantum mechanics been an advance or a retreat?

If it has been a retreat in any sense at all, it has been a

strategic retreat from the suffocating determinism of classical physics, which channeled and all but surrounded the advancing forces of science. Whether or not science, later in its quest, may once more encounter a deep causality, the determinism of the nineteenth century, for all the great discoveries it sired, was rapidly becoming an impediment to progress. When Planck first discovered the infinitesimal existence of the quantum, it seemed there could be no proper place for it anywhere in the whole broad domain of physical science. Yet in a brief quarter century, so powerful did it prove, it thrust itself into every nook and cranny, its influence growing to such undreamed-of proportions that the whole aspect of science was utterly transformed. With explosive violence it finally thrust through the restraining walls of determinism, releasing the pent-up forces of scientific progress to pour into the untouched fertile plains beyond, there to reap an untold harvest of discovery while still retaining the use of those splendid edifices it had created within the classical domain. The older theories were made more secure than ever, their triumphs unimpaired and their failures mitigated, for now their validity was established wherever the influence of the quantum might momentarily be neglected. Their failures were no longer disquieting perplexities which threatened to undermine the whole structure and bring it toppling down. With proper diagnosis the classical structures could be saved for special purposes, and their very weaknesses turned to good account as strong corroborations of the newer ideas; ideas which transcended the old without destroying their limited effectiveness.

True, the newer theory baffled the untutored imagination, and was formidably abstract as no physical theory had ever been before. But this was a small price to pay for its extraordinary accomplishments. Newton's theory too had once seemed almost incredible, as also had that of Maxwell, and strange though quantum mechanics might appear, it was firmly founded on fundamental experiment. Here at long last was a theory which could embrace that primitive, salient fact of our material universe, that simple, everyday fact on which the Maxwellian theory so spectacularly foundered, the enduring stability of the different elements and of their physical and chemical properties. Nor was the new theory too rigid in this regard, but could equally well embrace the fact of radioactive transformation. Here at last was a theory which could yield the precise details of the enormously intricate data of spectroscopy. The photoelectric effect and a host of kindred phenomena succumbed to the new ideas, as too did the wavelike interference effects which formerly seemed to contradict them. With the aid of relativity, the spin of the electron was incorporated with remarkable felicity and success. Pauli's exclusion principle took on a broader significance, and through it the science of chemistry acquired a new theoretical basis amounting almost to a new science, theoretical chemistry, capable of solving problems hitherto beyond the reach of the theorist. The theory of metallic magnetism was brilliantly transformed, and staggering difficulties in the theory of the flow of electricity through metals were removed as if by magic thanks to quantum mechanics, and especially to Pauli's exclusion principle. The atomic nucleus was to yield up invaluable secrets to the new quantum physics, as will be told; secrets which could not be revealed at all to the classical theory, since that theory was too primitive to comprehend them; secrets so abstruse they may not even be uttered except in quantum terms. Our understanding of the nature of the tremendous forces residing in the atomic nucleus, incomplete though it be, would be meager indeed without the quantum theory to guide our search and encourage our comprehension in these most intriguing and mysterious regions of the universe. This is no more than a glimpse of the unparalleled achievements of quantum mechanics. The wealth of accomplishment and corroborative evidence is simply staggering.

"Daddy, do scientists really know what they are talking about?"

To still an inquiring child one is sometimes driven to regrettable extremes. Was our affirmative answer honest in this particular instance?

Certainly it was honest enough in its context, immediately following the two other questions. But what of this same question now, standing alone? Do scientists really know what they are talking about?

If we allowed the poets and philosophers and priests to decide, they would assuredly decide, on lofty grounds, against the physicists—quite irrespective of quantum mechanics. But on sufficiently lofty grounds the poets, philosophers, and priests themselves may scarcely claim they know whereof they talk, and in some instances, far from lofty, science has caught both them and itself in outright error.

True, the universe is more than a collection of objective experimental data; more than the complexus of theories, abstractions, and special assumptions devised to hold the data together; more, indeed, than any construct modeled on this cold objectivity. For there is a deeper, more subjective world, a world of sensation and emotion, of aesthetic, moral, and religious values as yet beyond the grasp of objective science. And towering majestically over all, inscrutable and inescapable, is the awful mystery of Existence itself, to confound the mind with an eternal enigma.

But let us descend from these to more mundane levels, for then the quantum physicist may make a truly impressive case; a case, moreover, backed by innumerable interlocking experiments forming a proof of stupendous cogency. Where else could one find a proof so overwhelming? How could one doubt the validity of so victorious a system? Men are hanged on evidence which, by comparison, must seem small and inconsequential beyond measure. Surely, then, the quantum physicists know what they are talking about. Surely their present theories are proper theories of the workings of the universe. Surely physical nature cannot be markedly different from what has at last so painfully been revealed.

And yet, if this is our belief, surely our whole story has been told in vain. Here, for instance, is a confident utterance of the year 1889:

"The wave theory of light is from the point of view of human beings a certainty."

It was no irresponsible visionary who made this bold assertion, no fifth-rate incompetent whose views might be lightly laughed away. It was the very man whose classic experiments, more than those of any other, established the electrical character of the waves of light; none other than the great Heinrich Hertz himself, whose own seemingly incidental observation contained the seed from which there later was to spring the revitalized particle theory.

Did not the classical physicists point to overwhelming evidence in support of their theories, theories which now seem to us so incomplete and superficial? Did they not generally believe that physics was near its end, its main problems solved and its

basis fully revealed, with only a little sweeping up and polish-
ing left to occupy succeeding generations? And did they not
believe these things even while they were aware of such
unsolved puzzles as the violet catastrophe, and the photoelectric effect, and radioactive disintegration?

The experimental proofs of science are not ultimate proofs.
Experiment, that final arbiter of science, has something of the
aspect of an oracle, its precise factual pronouncements couched
in muffled language of deceptive import. While to Bohr such
a thing as the Balmer ladder meant orbits and jumps, to
Schrödinger it meant a smeared-out essence of ψ; neither view
is accepted at this moment. Even the measurement of the
speed of light in water, that seemingly clear-cut experiment
specifically conceived to decide between wave and particle,
yielded a truth whose import was misconstrued. Science
abounds with similar instances. Each change of theory demonstrates anew the uncertain certainty of experiment. One would
be bold indeed to assert that science at last has reached an
ultimate theory, that the quantum theory as we know it now
will survive with only superficial alteration. It may be so, but
we are unable to prove it, and certainly precedent would seem
to be against it. The quantum physicist does not know whether
he knows what he is talking about. But this at least he does
know, that his talk, however incorrect it may ultimately prove
to be, is at present immeasurably superior to that of his
classical forebears, and better founded in fact than ever before.
And that is surely something well worth knowing.

Never had fundamental science seen an era so explosively
triumphant. With such revolutionary concepts as relativity
and the quantum theory developing simultaneously, physics
experienced a turmoil of upheaval and transformation without

parallel in its history. The majestic motions of the heavens and the innermost tremblings of the atoms alike came under the searching scrutiny of the new theories. Man's concepts of time and space, of matter and radiation, energy, momentum, and causality, even of science and of the universe itself, all were transmuted under the electrifying impact of the double revolution. Here in our story we have followed the frenzied fortunes of the quantum during those fabulous years, from its first hesitant conception in the minds of gifted men, through precarious early years of infancy, to a temporary lodgment in the primitive theory of Bohr, there to prepare for a bewildering and spectacular leap into maturity that was to turn the orderly landscape of science into a scene of utmost confusion. Gradually, from the confusion we saw a new landscape emerge, barely recognizable, serene, and immeasurably extended, and once more orderly and neat as befits the landscape of science.

The new ideas, when first they came, were wholly repugnant to the older scientists whose minds were firmly set in traditional ways. In those days even the flexible minds of the younger men found them startling. Yet now the physicists of the new generation, like infants incomprehensibly enjoying their cod-liver oil, lap up these quantum ideas with hearty appetite, untroubled by the misgivings and gnawing doubts which so sorely plagued their elders. Thus to the already burdensome list of scientific corroborations and proofs may now be added this crowning testimony out of the mouths of babes and sucklings. The quantum has arrived. The tale is told. Let the final curtain fall.

But ere the curtain falls we of the audience thrust forward, not yet satisfied. We are not specialists in atomic physics. We are but plain men who daily go about our appointed tasks, and of an evening peer hesitantly over the shoulder of the scientific theorist to glimpse the enchanted pageant that passes before his mind. Is all this business of wavicles and lack of causality in space and time something which the theorist can now accept with serenity? Can we ourselves ever learn to welcome it with any deep feeling of acceptance? When so alien a world has been revealed to us we cannot but shrink from its vast unfriendliness. It is a world far removed from our everyday experience. It offers no simple comfort. It beckons us without warmth. We are saddened that science should have taken this curious, unhappy turn, ever away from the beliefs we most fondly cherish. Surely, we console ourselves, it is but a temporary aberration. Surely science will someday find the tenuous road back to normalcy, and ordinary men will once more understand its message, simple and clear, and untroubled by abstract paradox.

But we must remember that men have always felt thus when a bold new idea has arisen, be the idea right or wrong. When men first proclaimed the earth was not flat, did they not propose a paradox as devilish and devastating as any we have met in our tale of the quantum? How utterly fantastic must such a belief at first have appeared to most people; this belief which is now so readily and blindly accepted by children, against the clearest evidence of their immediate senses, that they are quick to ridicule the solitary crank who still may claim the earth is flat; their only concern, if any, is for the welfare of the poor people on the other side of this our round earth who, they so vividly reason, are fated to live out their lives walking on their heads. Let us pray that political wisdom and heaven-sent luck be granted us so that our children's children may be able as readily to accept the quantum horrors of today and laugh at the fears and misgivings of their benighted ancestors, those poor souls who still believed in old-fashioned waves and particles, and the necessity for national sovereignty, and all the other superstitions of an outworn age.

It is not on the basis of our routine feelings that we should try here to weigh the value and significance of the quantum revolution. It is rather on the basis of its innate logic.

"What!" you will exclaim. "Its innate logic? Surely that is the last thing we could grant it. We have to concede its overwhelming experimental support. But innate logic, a sort of aura to compel our belief, experiment or no experiment? No, that is too much. The new ideas are not innately acceptable, nor will talking ever make them so. Experiment forced them on us, but we cannot feel their inevitability. We accept them only laboriously, after much obstinate struggle. We shall never see their deeper meaning as in a flash of revelation. Though Nature be for them, our whole nature is against them. Innate logic? No! Just bitter medicine."

But there is yet a possibility. Perhaps there is after all some innate logic in the quantum theory. Perhaps we may yet see in it a profoundly simple revelation, by whose light the ideas of the older science may appear as laughable as the doctrine that the earth is flat. We have but to remind ourselves that our ideas of space and time came to us through our everyday experience and were gradually refined by the careful experiment of the scientist. As experiment became more precise, space and time began to assume a new aspect. Even the relatively superficial experiment of Michelson and Morley, back in 1887, ultimately led to the shattering of some of our concepts of space and time by the theory of relativity. Nowadays, through the deeper techniques of the modern physicist we find that

space and time as we know them so familiarly, and even space
and time as relativity knows them, simply do not fit the more
profound pattern of existence revealed by atomic experiment.

What, after all, are these mystic entities space and time?
We tend to take them for granted. We imagine space to be so
smooth and precise we can define within it such a thing as a
point—something having no size at all but only a continuing
location. Now, this is all very well in abstract thought. Indeed,
it seems almost an unavoidable necessity. Yet if we examine
it in the light of the quantum discoveries, do we not find the
beginning of a doubt? For how would we try to fix such a disembodied location in actual physical space as distinct from
the purely mental image of space we have within our minds?
What is the smallest, most delicate instrument we could use
in order to locate it? Certainly not our finger. That could
suffice to point out a house, or a pebble, or even, with difficulty,
a particular grain of sand. But for a point it is far too gross.

What of the point of a needle, then? Better. But far from
adequate. Look at the needle point under a microscope and the
reason is clear, for it there appears as a pitted, tortured landscape, shapeless and useless. What then? We must try smaller
and ever smaller, finer and ever finer indicators. But try as we
will we cannot continue indefinitely. The ultimate point will
always elude us. For in the end we shall come to such things as
individual electrons, or nuclei, or photons, and beyond these,
in the present state of science, we cannot go. What has
become, then, of our idea of the location of a point? Has it not
somehow dissolved away amid the swirling wavicles? True, we
have said that we may know the exact position of a wavicle if
we will sacrifice all knowledge of its motion. Yet even here
there happen to be theoretical reasons connected with Comp-

ton's experiment which limit the precision with which this position may be known. Even supposing the position could be known with the utmost exactitude, would we then have a point such as we have in mind? No. For a point has a continuing location, while our location would be evanescent. We would still have merely a sort of abstract wavicle rather than an abstract point. Whether we think of an electron as a wavicle, or whether we think of it as a particle buffeted by the photons under a Heisenberg microscope, we find that the physical notion of a precise, continuing location escapes us. Though we have reached the present theoretical limit of refinement we have not yet found location. Indeed, we seem to be further from it than when we so hopefully started out. Space is not so simple a concept as we had naïvely thought.

It is much as if we sought to observe a detail in a newspaper photograph. We look at the picture more closely but the tantalizing detail still escapes us. Annoyed, we bring a magnifying glass to bear upon it, and lo! our eager optimism is shattered. We find ourselves far worse off than before. What seemed to be an eye has now dissolved away into a meaningless jumble of splotches of black and white. The detail we had imagined simply was not there. Yet from a distance the picture still looks perfect.

Perhaps it is the same with space, and with time too. Instinctively we feel they have infinite detail. But when we bring to bear on them our most refined techniques of observation and precise measurement we find that the infinite detail we had imagined has somehow vanished away. It is not space and time that are basic, but the fundamental particles of matter or energy themselves. Without these we could not have formed even the picture we instinctively have of a smooth, unblemished, faultless, and infinitely detailed space and time. These electrons and the other fundamental particles, they do not exist in space and time. It is space and time that exist because of them. These particles—wavicles, as we must regard them if we wish to mix in our inappropriate, anthropomorphic fancies of space and time—these fundamental particles precede and transcend the concepts of space and time. They are deeper and more fundamental, more primitive and primordial. It is out of them in the untold aggregate that we build our spatial and temporal concepts, much as out of the multitude of seemingly haphazard dots and splotches of the newspaper photograph we build in our minds a smooth, unblemished portrait; much as from the swift succession of quite motionless pictures projected on a motion-picture screen we build in our minds the illusion of smooth, continuous motion.

Perhaps it is this which the quantum theory is striving to express. Perhaps it is this which makes it seem so paradoxical. If space and time are not the fundamental stuff of the universe but merely particular average, statistical effects of crowds of more fundamental entities lying deeper down, it is no longer strange that these fundamental entities, when imagined as existing in space and time, should exhibit such ill-matched properties as those of wave and particle. There may, after all, be some innate logic in the paradoxes of quantum physics.

This idea of average effects which do not belong to the individual is nothing new to science. Temperature, so real and definite that we can read it with a simple thermometer, is merely a statistical effect of chaotic molecular motions. Nor are we at all troubled that it should be so. The air pressure in our automobile tires is but the statistical effect of a ceaseless bombardment by tireless air molecules. A single molecule has neither temperature nor pressure in any ordinary sense of those terms. Ordinary temperature and pressure are crowd effects. When we try to examine them too closely, by observing an individual molecule, they simply vanish away. Take the smooth flow of water. It too vanishes away when we examine a single water molecule. It is no more than a potent myth created out of the myriad motions of water molecules in enormous numbers.

So too may it well be with space and time themselves, though this is something far more difficult to imagine even tentatively. As the individual water molecules lack the every-day qualities of temperature, pressure, and fluidity, as single letters of the alphabet lack the quality of poetry, so perhaps may the fundamental particles of the universe individually lack the quality of existing in space and time; the very space and time which the particles themselves, in the enormous aggregate, falsely present to us as entities so pre-eminently fundamental we can hardly conceive of any existence at all without them. See how it all fits in now. The quantum paradoxes are of our own making, for we have tried to follow the motions of individual particles through space and time, while all along these individual particles have no existence in space and time. It is space and time that exist through the particles. An individual particle is not in two places at once. It is in no place at all. Would we feel amazed and upset that a thought could be in two places at once? A thought, if we imagine it as something outside our brain, has no quality of location. If we did wish to locate it hypothetically, for any particular reason, we would expect it to transcend the ordinary limitations of space and time. It is only because we have all along regarded matter as existing in space and time that we find it so hard to renounce this idea for the individual particles. But once we do renounce it the paradoxes vanish away and the message of the quantum suddenly becomes clear: space and time are not fundamental.

Speculation? Certainly. But so is all theorizing. While nothing so drastic has yet been really incorporated into the mathematical fabric of quantum mechanics, this may well be because of the formidable technical and emotional problems involved. Meanwhile quantum theorists find themselves more and more strongly thrust toward some such speculation. It would solve so many problems. But nobody knows how to set about giving it proper mathematical expression. If something such as this shall prove to be the true nature of space and time, then relativity and the quantum theory as they now stand would appear to be quite irreconcilable. For relativity, as a field theory, must look on space and time as basic entities, while the quantum theory, for all its present technical inability to emancipate itself from the space-time tyranny, tends very strongly against that view. Yet there is a deal of truth in both relativity and the present quantum theory, and neither can wholly succumb to the other. Where the two theories meet there is a vital ferment. A process of cross-fertilization is under way. Out of it someday will spring a new and far more potent theory, bearing hereditary traces of its two illustrious ancestors, which will ultimately fall heir to all their rich possessions and spread itself to bring their separate domains under a single rule. What will then survive of our present ideas no one can say. Already we have seen waves and particles and causality and space and time all undermined. Let us hasten to bring the curtain down in a rush lest something really serious should happen.

EPILOGUE

THOUGH the curtain has fallen, it must rise once more, for ours is a living story that will not rest. Two fateful decades in the affairs of man have passed since the climactic days of the quantum revolution. Scientists we have met have become political exiles far from the lands of their birth, symptoms of a malignant cancer whose baleful remnant lingers even yet. War again has smeared its crimson stain across the world; war in which the insubstantial equations of Clerk Maxwell furthered radar even as relativity and quantum mechanics aided the stupendous development of the atomic bomb.

Much has occurred in physical science since the stirring days of the quantum revolution. But with the enthronement of quantum mechanics the period of turmoil ended and subsequent events, though often unexpected, have at least been relatively orderly. The advent of the new theory meant a tremendous release. Obstructions to scientific progress which had persisted for decades, even centuries, were swiftly swept aside, and science leaped forward with renewed impetus. Under its leadership physics invaded vast new territories. Even the private preserves of chemistry were encroached upon, while the science of spectroscopy, which had played so brilliant and decisive a part in fostering the quantum mechanical revolution, was virtually overwhelmed, its many long-standing puzzles yielding impressive confirmations of the validity of the new theory.

With spectroscopy thus temporarily bereft of mystery, scientific theorists sought out deeper problems even as the stream of experimental research converged ever more strongly on those great enigmas, the atomic nucleus and cosmic rays.

Nuclear physics, which will forever be associated with the name of Rutherford, existed even before the nucleus was recognized, for observation of the radioactive process established many important facts now known to pertain to the nucleus. It was Rutherford who realized that the swift particles ejected by radioactive substances could probe for him the constitution of other atoms. It was Rutherford who, from the experiments of various physicists with these atomic projectiles, extracted a nuclear model of the atom the profound effect of which on physical theory we have here partially traced. And it was Rutherford who opened up a new vista for nuclear science in 1919 by his discovery of artificial nuclear transformation. For in that year he proved that, occasionally, when a fast α particle struck a nitrogen atom, a proton, or hydrogen nucleus, would be ejected. Soon other such artificial disintegrations were detected, and the chase was on in earnest. Rutherford it was who clamored for more powerful atomic projectiles with which to bombard the nucleus; but for long years man could not better the projectiles found in nature. Then in the early nineteen-thirties physicists began to devise highly ingenious machines for powerfully accelerating atomic particles, the most significant being the cyclotron, for the invention of which the American physicist E. O. Lawrence received the Nobel prize in 1939. With such machines as these to hand, artificial nuclear transformations became almost a commonplace. Yet some of the most important nuclear discoveries, of which we shall tell, were made with the natural projectiles of radioactivity and the cosmic rays.

With the spectacular advance of nuclear research, quantum mechanics was faced with a crucial test. For quantum mechanics was born of photons and electrons. Spectroscopy, the photoelectric effect, and the other regions of its splendid triumphs lay almost wholly within the domain of the electron and the photon. Now, with men exploring the unknown realms of the nucleus, it was to be called upon to pit its strength against experimental discoveries on a far deeper plane. Here was the supreme test of the fledgling powers of the youthful quantum mechanics. Would this latest theory, sired, nurtured, and lovingly coddled by the photon and electron, successfully encompass the new discoveries in these pioneering realms of scientific exploration, or would it on its first external test in the harsh world away from home exhibit, like so many theories before it, serious and perhaps fatal deficiencies? The time had come for quantum mechanics to stand on its own and find its rightful place in the rapidly widening world of physical discovery. Would that place be high or low, long-lived or transitory?

It was in the year 1928, early in the life of the new theory, that J. R. Oppenheimer noticed a certain mathematical peculiarity of quantum mechanics. Shortly thereafter the theoretical physicist G. Gamow made use of this peculiarity in a significant application to the nuclear problem of radioactivity. So early was it in the life of the new theory, indeed, that the quantum had not yet outgrown its infantile pranks, for Gamow's discovery was made independently, at practically the same time, by the English and American collaborators R. W. Gurney and E. U. Condon.

There had long been a baffling discrepancy about radio-activity. It was a discrepancy easy to state and simple to understand; and paradoxically its very simplicity made it all the more baffling, for had it been complex one might have sought some subtle, intricate loophole, but against so bald and uncompromising a discrepancy subtlety seemed powerless. Here was the problem: An atomic nucleus could be conceived as a sort of volcano within whose crater seethed a restless ocean of particles. An ordinary nucleus would correspond to an extinct volcano, a radioactive nucleus to an active one. If a minor volcanic eruption occurred and a particle was ejected from the crater, that would correspond to the emission of a particle by a radioactive nucleus. This was about the only picture of a radioactive nucleus available. It worked by no means badly.

But alas it conflicted with experiment. If particles came over the top of the crater, they should fall outside with considerable speed. Measurements showed the particles did not have the required speed. That was all. A mere discrepancy. But even commercial banks stand aghast at a discrepancy. If the nuclear energy books did not balance, then something must be wrong. There was nothing intricate about the energy accounting. The discrepancy could not be due to some subtle falsification. It was as barefaced a discrepancy as one could possibly imagine, and in banking circles would surely send someone to jail. No wonder, then, that in scientific circles it caused considerable perturbation. For if the measurements were correct, and no one pretended they were not, the particles could not possibly have come over the top of the crater. How, then, was the Houdini-like escape effected? It looked as if the convenient volcano picture must be basically incorrect. Only something of a miracle could save it.

It was here that Gamow brought in quantum mechanics to offer its unique services. Was a miracle needed? Then quantum mechanics would provide one. But a logical one—according to the curious logic of quantum happenings. Instead of imagining the volcano to be a teeming hive of scurrying particles, we must imagine within its crater a vibrant, surging lava of ψ probability. That much is quantum mechanically reasonable, surely. And now the mathematics shows that, somewhat as sound waves issue from a closed room, so do the ψ vibrations insinuate themselves through the crater walls and set up ψ waves outside. This pronouncement of the mathematics, though strange, was very welcome. For what were these ψ waves which had thus managed to appear outside the confines of the crater's walls? According to quantum mechanics, were they not probabilities? And what sort of probabilities? Why, probabilities that a particle was actually there somewhere outside the crater, having mysteriously passed through the volcanic walls with never a hole or blemish to mark its passage.

Could anything but quantum mechanics have rescued the volcano picture so audaciously? There was indeed more than audacity in the rescue operation. It was a rescue in the grand manner, with not even a trace of niggardliness. For in removing the primary difficulty of the speeds it also accounted for a variety of well-known characteristics of radioactivity, such as the experimentally discovered relation between the speeds and the rate of decay, and that paramount fact of nature, the existence of a ladder of energy levels within the nucleus.

Such was the first attack upon the nucleus, and surely it was a major triumph. Yet it was no more than a preliminary skirmish. The nucleus was not to be vanquished so easily. It had not begun to yield its deeper secrets. The volcano picture was still beset with difficulties, and woefully lacked the detail needed for an adequate concept of the nucleus.

One thing was particularly disquieting about the volcano picture. When dealing with the helium particles shot out by a radioactive nucleus it gave the splendid results recounted above. But for electrons shot from the nucleus it gave the same sort of results. That does not seem cause for concern. What's sauce for the helium particles is surely sauce for the electrons, isn't it? One would certainly think so. But the electrons had other views. They came out of the nucleus with speeds which flatly contradicted the idea of an energy ladder within. Nor was this all. For, speaking loosely, the electron was found to be larger than the nucleus. How it could ever get inside was thus a major mystery—even for quantum mechanics. There was something decidedly queer about the electron. The quantum theory almost came a cropper over it. How it ultimately contrived to save itself, and with what brilliant adroitness and intricacy of maneuver, is a truly fascinating story. A humane author would hasten to tell it at once. Let me tell you instead of the curious incident of the Dirac protons.

Two chapters ago Dirac had just succeeded in wedding relativity to quantum mechanics, the electron spin emerging as offspring. Dirac's equations were undoubtedly well fitted to describe the behavior of electrons. They did so superbly, with every little detail faithfully delineated. Unfortunately, they did more than seemed really necessary, not only describing the usual behavior of an electron but also another mode of behavior in which it had a negative amount of energy, which is to say a negative weight. If pushed by a force acting toward the right, it would move toward the left. This, of course, was nonsense, and it would have been natural enough to ignore it as an unfortunate but luckily unimportant idiosyncrasy of otherwise excellent equations. But the negative energy states of the electrons could not be ignored.

In the prequantum era all would have been well, for there was a gap between the negative and positive states that could not be bridged classically. But alas we are in the quantum age and our electrons will jump from one energy value to another, for that is the hallmark of the quantum regime. And see what that would mean. If we started with an ordinary electron, it could jump and become a nonsensical one of negative energy. Dirac was greatly perturbed. There was nothing in the equations to prevent these jumps. Either there was a grave defect in his theory or else it was trying to convey an urgent message. What could that message be? An electron liable at any moment to change into the scientific equivalent of a mythological monster is surely no electron for the serious scientist. Yet there was nothing in the equations that could possibly prevent this Jekyll-Hyde transformation.

Nothing in the equations—but how about something outside? For example, how about the Pauli exclusion principle, that rigid ordinance against overcrowding? If we drag that in we may possibly save the situation. It is a desperate remedy, but it may work. And if it does work it will be well worth the effort. Suppose we imagine all the states of negative energy already occupied by these mythical monsters. Then the ordinary electrons may no longer jump into these states. The exclusion principle prevents them. The jumps would create overcrowding. Our problem is solved. We can ignore the jumps, after all. And it's really not too bad a solution at that.

But alas our problem is not solved. The idea will not work. The remedy turns out to be just as bad as the disease. To be sure, the ordinary electrons can no longer jump into the negative energy states. But what is to prevent these myriads of negative energy electrons from jumping into the positive energy states and thus suddenly changing from lazy monsters to well-behaved electrons? Once this happens, the jumps can even proceed in both directions. It's a pity it would not work. So heroic a measure as introducing multitudes of lethargic monsters deserved a better fate. There is no sense wasting time in idle lamentation, though. Work is to be done. We still have our problem to solve.

What is another possible line of attack? Can we perhaps alter the equations just a little? Put in some extra mathematical term, an x here or a y there? It will be hard to do without damaging the equations, but at least it is a possibility. No harm in trying. Let us see. It is not going to be too easy. There's relativity to consider too, and that makes changes difficult. We would have to be careful to put in the extra bit of mathematics where it would not interfere with the . . . But wait! Quick! Back to the exclusion principle. It's going to be all right, after all. We can save everything. And get something very wonderful out of it too. Look. Suppose we did have all the negative energy states full, a veritable ocean of mythical monster electrons. And suppose that then one of these monsters did suddenly jump to a positive energy state. An ordinary electron would suddenly appear. But there would also be a tiny bubble in the monstrous ocean where the monster suddenly ceased to be. Never mind the ordinary electron for the moment. Concentrate on the bubble. How will it look? How will it behave? We must do a little calculation of course, but it is easily performed. Here it is. Yes. The bubble, being, so to speak, an absence of a negative energy, will behave as if it has a positive energy. That means it will seem like an ordinary well-behaved particle. Good. Splendid! Will it seem like an ordinary electron, then? Let us have another look at the equations. Will it? No. It will move in the direction opposite to that expected; that means it will have just the opposite electric charge. What has just the opposite charge to an electron? Why, a proton of course. What a wonderful chance to create a theory of protons as a mere by-product of our theory of electrons. That would be really something; a grand unification if ever there was one; one of those big things in scientific theory that come once in a lifetime—maybe twice for a Dirac. We must certainly work on it some more. How heavy will the new bubble-particle seem to be? It is easily calculated. The mass will be the same as that of an electron.

But that is very bad. It does not fit. A proton is almost two thousand times as heavy as an electron. If we are on the right track, whence comes this enormous mass? Perhaps we can unearth some unexpected idiosyncrasy in the electron equations, some undiscovered lack of symmetry. Perhaps the presence of all those low-energy monsters might have a biasing effect upon the masses. It is worth a tremendous effort to find out. The prize is dazzling.

Dirac had gaily gone so far. And then his luck gave out. By no amount of effort could he make the mass come out right. It clung tenaciously to the electronic value, simply refusing to budge, and Dirac had on his hands a bitterly disappointing theory of electrons and protons which maintained that the masses of proton and electron were equal—off by a mere two hundred thousand per cent. There was nothing for it but to admit defeat and announce the theory in this deplorable condition in the hope that other, fresher minds might find a way to patch it up.

The months rolled by, and the years. But where Dirac had failed, no one else succeeded. The mass remained obdurate. The theory was plainly defective, and Dirac no longer referred to his mythical monsters as protons but called them antielectrons.

And then, some four years later, in that magic year 1932, the young American experimenter Carl D. Anderson performed the experiments for which, in 1936, he was to receive the Nobel prize. Our story has been progressing toward ever more comprehensive unifications. Now it suddenly takes on a different aspect. While investigating the effects of cosmic rays, Anderson discovered a new type of particle, the positron. Despite experimental difficulties, for the positron is an elusive, short-lived particle, the evidence indicated strongly that positron and electron have opposite electrical charges and equal masses. Here was Dirac's "proton"; here was his antielectron. The theory of the monstrous bubbles was vindicated.

There was more to Anderson's discovery than this, though, something really exciting. Photons of enormous energy in the cosmic rays were being transmuted into pairs of electrons and positrons. Radiation—light—was changing into matter in accordance with Einstein's famous law of the equivalence of mass and energy. In the restless underworld of our material universe this tremendous process of seething transformation had been going on since time immemorial.

Does the transformation of light into matter seem almost impossible to visualize? Let us look at it with the aid of Dirac's theory. Light of enormous energy, striking down into the murky ocean, is swallowed up by a lethargic monster. The monster, full of energy after its light meal, immediately leaps from the ocean, changing thereby into a regular electron, and leaving a tiny bubble where it had been. This bubble is a positron, inescapable companion of the newly created electron. If later the electron wishes to return to its earlier monstrous state it must disgorge its surplus energy and rejoin its former companions in the ocean, filling the waiting bubble. To us this would appear as if an electron and a positron had suddenly crashed head-on and vanished amid a burst of radiation—matter transmuted back into energy. Since this reverse process limits the average life span of the positron to an incredibly small fraction of a second, it is perhaps no wonder that the positron escaped our observation all these years. Yet once attention had been directed toward it, it was readily detected by many observers, and in 1933 was found to play an important nuclear role. For in that year the French scientist F. Joliot and his wife I. Curie, daughter of Marie Curie, discovered that radioactivity could be induced in certain nuclei by a bombardment of α particles, and found too that positrons were shot out by nuclei thus rendered artificially radioactive. For this discovery, so happily in the family tradition, they received the Nobel prize for chemistry in 1935.

Does the discovery of the positron suffice to make a magic year? Perhaps. But surely no more magical than many another year. There was more that happened in 1932. And certainly it was magical, for how else may we explain the fascinating pattern which the strange forces of coincidence fashioned for our delight? It was a year consecrated to the experimenter, a fabulous, phenomenal year which brought to light as many as three new particles. The positron was but one of a trilogy. Two other major discoveries came that year. There was, for instance, the discovery of heavy water by the American chemist H. C. Urey, which won for him the Nobel prize just two years later. What made the water heavy was the presence of heavy hydrogen, whose atomic nucleus, a particle twice as massive as the proton but of equal charge, was something hitherto quite unknown to science. When finally this new particle was christened deuteron, a wit remarked that Urey had created the new science of deuteronomy.

Though the deuteron is of major significance, and though heavy water was so potentially important for atomic bomb research that men gave their lives to prevent the Nazis from using it, it is nevertheless of relatively minor interest for our story of the quantum. Its discovery was startling, to be sure, and, as we shall see, it did point up certain defects in previous ideas about the nucleus. But it had none of the impact of that other event of 1932, that epochal event which was to revolutionize our theories of the nucleus, and without which atomic bombs would still be mercifully denied us, the discovery by the English physicist J. Chadwick of yet another new type of particle, the neutron. For his world-shattering discovery—we use the phrase not lightly—Chadwick received the Nobel prize in 1935. The possible existence of the neutron had been conjectured in 1920 by Rutherford, and simultaneously by the American chemist W. D. Harkins, on the basis of the properties of nuclei. On the experimental side, too, Chadwick's work was but the brilliant culmination of the pioneering researches of the German physicists W. Bothe and H. Becker and the further investigations of Curie and Joliot.

The neutron is a particle about as heavy as the proton, but without electrical charge. We know that a hydrogen atom consists of a proton and an electron. What if the electron should fall into the proton? There might be quite an explosion. But suppose it were more peaceful, so that proton and electron simply merged. There would then result an electrically neutral particle. It was this possible particle, this collapsed hydrogen atom, which Rutherford and Harkins had envisaged. Pauli, too, had felt the need for some sort of neutral particle in order to explain an anomaly in the nucleus of the nitrogen atom. But all this was far from direct experimental demonstration. When Bothe and Becker noticed a curious radiation which seemed to defy the accepted laws of physics, the further investigations of the Curie-Joliots served but to emphasize its strange behavior. Chadwick, a former pupil of Rutherford, brilliantly put two and two together and demonstrated conclusively, both experimentally and theoretically, that this new radiation must consist of the uncharged particles we call neutrons, a name already used by Harkins and Rutherford a dozen years before. Were this the story of the development of physics, or even the oft-told tale of the atomic bomb, we could pause to tell the many interesting details of this quest, how the great Italian scientist Enrico Fermi used neutrons to bombard uranium, and with what curious results, and how this ultimately led to the discovery of nuclear fission, with all that that entails. But this is the story of the quantum. We dare not stray too far from our central theme.

For all that the existence of the neutron had been anticipated, its advent found the theorists unprepared, their
theories unable to encompass it. The idea of treating it as a
collapsed hydrogen atom proved unsuccessful. Now that the
neutron had been detected by the experimenters, the evidence
began to pour in to confirm that the neutron was no such
naïve combination of well-known particles. It was no im-
poverished beggar at the feast, but a high dignitary in the
mansions of science to be treated with every proper mark of
respect and accepted in its own proud right as a fundamental
particle.

For the theorists all this seemed a disquieting retrogression.
Hitherto they had taken this complex, puzzling world of ours
and, with a grand feeling of heart-warming satisfaction, reduced it to nothing but protons, electrons, and photons.

Had Dirac succeeded with his theory of the proton, what a
magnificent unification would have ensued; the whole universe built of nothing but electrons and photons. Alas, the
positron spoiled such lofty dreams. And the neutron, coming
at practically the same time, seemed to tear their gossamer
fabric into shreds.

Somehow the shattered dreams must be replaced. This was
no time for despair. A challenge had to be met. Though neutrons might not be the collapsed hydrogen atoms of Harkins
and Rutherford, if they had been knocked out of the nucleus
a place must somehow be found for them inside. It was still
the year 1932 when Heisenberg made the first successful
theoretical attack on the detailed internal structure of the
nucleus. In it he assumed that nuclei contain protons and
neutrons only. How radical a departure this was from previous

ideas, how great its revolutionary import, how daring its implications, events will show.

Had you by any chance forgotten about the electrons, and how the volcano picture could not handle them? It could not handle the positrons either. If the nucleus contains only protons and neutrons, one of our nuclear problems is solved right away. We need worry no more that electrons and positrons are too large. Neutrons and protons are very much smaller. For them there is room enough within the nucleus. In fact, they fit quite snugly. And if we use only protons and neutrons we can overcome some hitherto puzzling discrepancies. For there are many numerical facts about the various nuclei that must be correctly accounted for: their mass, and charge, and spin, and other things.

The old idea could explain practically all of these, but in the case of nitrogen there was a discrepancy. For the nitrogen nucleus, which has a mass fourteen times that of the proton, must, on the old view, contain fourteen protons. Since its charge is only seven times that of the proton, it must contain seven electrons to neutralize the excess charge of the fourteen protons. Thus mass and charge have told precisely how many protons and electrons must be contained. The spins must therefore balance automatically. Now, the spins of protons and electrons are each known to be half a unit. Their values may be combined, or they may cancel in pairs. But with twenty-one particles in all, which is an odd number, it is impossible to obtain a whole-number spin; and unfortunately the spin of the nitrogen nucleus in question is one whole unit. With neutrons and protons this can at once be accounted for if neutrons, like protons, have spin of half a unit. For now the nitrogen nucleus will consist of seven protons and seven neutrons, an even number of particles in all. Let six of these cancel the spins of six others, and the remaining two combine their half-unit spins, and the total spin comes out to be one unit as required. The newly discovered deuteron was later found to provide a similar corroboration of the new idea. Since the deuteron has twice the mass of the proton and a charge equal to the proton charge, the old view would require it to consist of two protons and an electron. But its spin was found to be one unit, and one unit could not arise from an odd number of protons and electrons. The new idea encounters no such difficulty, for it would have the deuteron consist of one proton and one neutron.

Yes, it does sound rather good. But does it really solve our problem? Does it not rather exchange one worry for another? If there are no electrons or positrons inside the nucleus, pray how does it happen they come shooting out?

Now we shall see what devious cunning pervades our nuclear theories. It is subtle with all the twisted subtlety of the quantum; strange and topsy-turvy, and yet for us who are now so deeply steeped in quantum lore, so utterly logical. For it is rooted in remembrance of things past. The mental path is clear. We know well how the scientists came upon their idea. Let us follow the path their thoughts once trod.

Ages ago, as it must seem to us now, though only a few short years in actual time, young pioneering Bohr had pictured the atom as a nucleus surrounded by electrons jumping from orbit to orbit. Despite the happenings of the intervening years, age has not marred the vivid potency of this rough picture. It remains a valued guide and counselor. Let us ask it a decisive question, a question not yet about protons and neutrons, nor yet about electrons and positrons, but rather about those swift, ethereal particles the photons. What happens precisely to a photon when it is swallowed up by an atom?

It is a question easily asked. Why, then, does the oracle remain silent? It is a silence of wisdom, an oracular answer whose message we must ponder, a pregnant silence from which will grow great things.

When a photon of the proper energy is so unfortunate as to strike a Bohr atom, that is the end of its individuality as a photon. It vanishes completely away, and in vanishing causes an electron to jump from one orbit to another of higher energy. Could one really say that the photon is trapped within the atom? No one would ever recognize it in captivity. It is changed into something utterly different—an electron jump which formerly occurred.

There is the reverse process too. What happens when an electron jumps to an orbit of lower energy? A photon magically appears as if from nowhere. There had been no photon within the atom. The newly created photon was but an external symptom of the electron's convulsive jump.

At first this hide-and-seek game of hunt the photon caused great concern. Men asked what happened while the electron was in the act of jumping, and by just what wizardry the photon vanished away into nothingness, its place to be taken by a mere colorless jump in energy. But the deeper maturity that came with quantum mechanics taught us that just such questions must forever remain unanswered, the alleged happenings to which they have reference being veiled by the indeterminacy principle.

But indeterminacy principle or no, the photon remained a very curious particle, quite different from the electron in a most important particular. Electrons, like protons, were indestructible, admirable building material for an imperishable universe. But photons! Why, photons were mere will-o'-the-wisps, evanescent and insubstantial, their energy alone abiding. True, they behaved like wavicles, just as the electrons did, but they were free to come and go; to come out of nothingness and return to nothingness; to materialize as radiant, lustrous wavicles and melt away again into black, lightless energy jumps. There was nothing solid or stolid about them. They had no continuing personality. They were Protean rather than protonic. They could multiply like rabbits. You could never be sure how many you had. You might even start with none at all and suddenly find yourself overwhelmed by them. Is not that precisely what happens when an atomic bomb explodes, or even an ordinary bomb though less spectacularly? In an instant, along with other effects by no means negligible, there appears a stupendous plenitude of photons, a dazzling flash of light where previously all was darkness, bright photons brought suddenly into existence in numbers of staggering splendor. Nor do we need a bomb to effect such creation. It is a commonplace. We do it every day without thinking by simply pressing a switch to turn on the light.

This flighty propensity of the photon of jumping into and out of existence sets it apart from the sturdy, reliable electron. Who could have foreseen that the seemingly imperishable electron was destined to go the way of the wayward photon? Yet the oracle taught us well by its silence. And did not the coming of the positron destroy the vaunted claims of the electron to imperishability? Like the photons, electrons and positrons jump into and out of existence. Their ignominy is truly complete, for they even change into the very photons they once pretended to despise for being so evanescent.

If electrons and positrons are so like the photon in this particular, why should they not also behave like photons in other respects? If they are too big for the nucleus, why could they not be definite personalities outside the nucleus but mere unrecognizable jumps of state within? This would preserve all the triumphs of the prenuclear era of quantum physics while saving the quantum from serious nuclear embarrassment. Indeed, there are many important details about the nucleus which we have not mentioned at all, which indicate that had such an idea not been envisaged the nucleus would have utterly vanquished the quantum. It was a narrow escape. But the way of escape was strictly within the established traditions of the quantum way of thought. No longer do we believe that the nucleus is built of protons and electrons. We think of it now as made up of protons and neutrons. Electrons and positrons are never inside the nucleus. They are external manifestations of jumps occurring within.

This is a novel concept. And it has a novel implication, for it means that the electrically charged proton and the electrically neutral neutron are actually one and the same particle.

Of course, this seems absurd. How can the uncharged neutron be the same particle as the charged proton?

The oracle has already supplied the answer. What happens when a photon merges with an electron? The electron jumps and the photon is no more. Electron-plus-photon is not some new composite particle. It is just the same old electron as before but in a new state of energy. With the nucleus made up of protons and neutrons, what would be the analogous thing? When an electron merged with a proton there would be a jump in the proton's state of energy and charge, and the electron would be no more. Proton-plus-electron would not be some new composite particle. It would be just the same old proton as before, but in a new state of energy and charge; the same old proton but now electrically neutral, for the electronic and protonic charges exactly balance; the same old proton, but we would call it now a neutron. And if a positron should similarly merge with a neutron, or a neutron should create and shed an electron, the neutron would jump back to its old protonic state.

That is how we must think of the nucleus. That is the stuff of which the uranium and plutonium of atomic bombs are made. That is the fanciful way we must build our universe. Though the concept is subtle and tenuous, the analogy with the photon is perfect, even to the detailed mathematical treatment.

Our story still has surprises. Have you by any chance forgotten about the electrons? Though we have now banished them from the nucleus, they are still able to get into mischief. Remember, they still contradict the incontrovertible fact that there is an energy ladder within the nucleus. Somehow the energy books do not balance. Can the quantum help us out once more? Only by emphasizing that the spin accounts also do not balance when, for instance, a proton turns into a neutron or vice versa. This was enough for Pauli, though. If the books do not balance, there must be a thief at work, said he. A new type of particle must be declared to exist, a marauder which steals off with some of the energy and spin and leaves no trace. But how can so bold a particle escape observation? Clearly we must endow it with a cloak of invisibility. Divest it, then, of telltale electric charge. Allow it only the most minute mass. If necessary, let it have no mass at all. Such a disembodied particle could pilfer right under our nose and escape detection; its nefarious activities would come to light only through an auditing of the electron accounts. And what could be a more fitting name for this diminutive neutron than the Italian diminutive *neutrino*?

Our sole clue to the character of the neutrino would be the character of its thefts, for the thief would never be caught. Clearly, on the circumstantial evidence, a case could be made against it. But would the case hold water? In the hands of Fermi and his followers the idea of the neutrino was developed into a full-fledged mathematical theory. Everything hinged on the consistency of the evidence when subjected to the rigors of a searching cross-examination of a profundity and intensity such as only a powerful mathematician could conceive. Despite some difficulties still not fully resolved, the available evidence was found to present a reasonably consistent picture of the invisible thief, and the marauding neutrino was accordingly admitted to the sacred halls of science. It was born in 1933, almost within the magical year of the particles.

Through this bit of detective work hitherto unsolved mysteries were now cleared up. Whenever a proton changed into a neutron, or a neutron into a proton, a neutrino must be involved along with the positron or electron, lest the ledgers betray unaccountable deficiencies.

But though the neutrino had joined the merry throng, there was yet a mystery to be solved, yet a discrepancy to thrust its stern compulsion on men's thoughts.

Within the nucleus are stupendous forces of fabulous

power welding its separate parts into a compact whole. From elementary characteristics of the different nuclei it could be seen that these forces must be analogous to the chemical forces that bind atoms into molecules. Heisenberg therefore ascribed them to exchange phenomena within the nucleus, his scheme being later modified in an important detail by E. Majorana. According to the curious picture scientists have to use in thinking of these things, this exchange is a sort of rhythmic interchange of position between the particles comprising the nucleus.

Now a neutron can become a proton by shedding an electron and a neutrino, and a proton can become a neutron by absorbing them. Thus the interchange of place between a proton and a neutron can be pictured as a sort of tossing to and fro between them of an electron and a neutrino, as in a long, fast rally in tennis. The neutron serves, and in serving becomes a proton. The original proton receives, and in receiving becomes a neutron. It at once returns the serve, and so reverts to its proton state while converting its opponent back into the neutron state. The effect of such a rally is a rhythmic alternation in which at one moment we have a neutron on this side of the net and a proton on that, the next moment a proton here and a neutron there, and so on back and forth.

If we wish to picture in this way how two neutrons could exchange places, or two protons, we would have to imagine a tennis game played with two balls at once flying in opposite directions.

Of course we may not think of this travel to and fro too literally. After all, the various "particles" involved are all wavicles, and so far as it is permissible to talk of their sizes at all, the electron would be larger than the proton and the neutron, and the whole nuclear tennis court. Thus surrounding the neutrons and protons of the nucleus is a ghostly halo of electrons and neutrinos fluctuating uncertainly between existence and nonexistence. This electrical halo of wavicles is linked with the electromagnetic field that Maxwell had conceived so many years ago as the seat of Faraday's tubes of force. And through his tennis-rally mechanism Heisenberg sought to establish a deep connection between it and the gigantic forces within the nucleus.

The idea was attractive but alas for all its undeniable charm it would not quite work. Though forces could be deduced from it, and enormous, fabulous forces at that, the theorists, like a sleeper on a wintry night whose blanket is too short, found themselves involved in a hopeless dilemma. It was no problem at all for them to give the forces the proper energy content, but if they did the forces would not reach one two-hundredth far enough. True, the theorists could stretch the reach of the forces, but that was like pulling a blanket up to the ears only to expose the toes. Worse even. For if they extended the reach of the forces the necessary two hundred and fiftyfold, the energy content became not just hundreds but hundreds of billions of times too weak.

Either the toes or the ears could be kept from freezing, but not both. The infant theory seemed doomed to die of lack of energy brought on by unavoidable exposure. It was rescued from untimely death by the Japanese scientist H. Yukawa, who found a theoretical way out of the impasse; a way to cover both ears and toes simultaneously. It involved something then fast becoming a habit among physicists; almost an occupational disease of the mind. Can you guess? What was the prevailing fashion in those days? New particles, was it not? Positrons. Neutrons. Neutrinos. And now yet another. For Yukawa proved that the nuclear forces could be made to fit as Heisenberg had originally hoped if only we would imagine a new type of wavicle in that powerful halo surrounding the protons and neutrons. The tennis must be played with a new type of ball. Not only was its mass to be intermediate between that of the neutron and that of the electron, it was to be an actual intermediary between the two. For, according to Yukawa, when a neutron changed into a proton and emitted an electron and a neutrino from a nucleus during radioactivity, it did not at once create and shed the electron and neutrino, as formerly thought. It first created a Yukawa particle, which, after an incredibly brief life span, exploded into two fragments that were the electron and neutrino of the older idea.

The pace of discovery was swift, Heisenberg mentioning his idea in 1934 and Yukawa proposing the new particle in 1935, a purely theoretical speculation, interesting and promising, but unconfirmed. Confirmation was not long delayed, however, for as early as 1936 the new type of particle was actually observed among the cosmic rays, one of the first to notice it being the same Anderson who discovered the positron. To resolve some of the many grave difficulties that still remain in the theory of nuclear forces, it has been necessary to assume various different types of Yukawa particles, both charged and electrically neutral. Such particles are now experimentally recognized as important constituents of the cosmic rays, yet there is still no agreement as to their name, some physicists referring to them as mesotrons, while others call them mesons.

The equations governing the mesons ultimately proved

familiar, bearing a striking resemblance to the equations which Maxwell had given for the electromagnetic field. One could take the equations of Maxwell and by a quite small alteration convert them into the meson equations. Could it be that the quantum physicists were entering a second childhood? Could it be that there was life in the older theories yet?

From our present state of knowledge, with our positrons and neutrinos, our neutrons and our various types of mesons, how sketchy and primitive the old Rutherford-Bohr atom now appears, and how extraordinarily effective considering its utter crudity. Surely in years to come men will look back on our present tentative gropings with the same wonder and tolerant admiration, amazed that concepts so crudely incomplete and incorrect should nevertheless probe so deep, and wrest from Mother Nature so many precious, dark, and terrible secrets, and expose our infantile civilization to such horrible dangers.

Though the picture of the nucleus grew more detailed, increasing knowledge did not smooth the path of the theorist. The nucleus was too complex a structure. It had too many particles in close proximity to yield readily to detailed mathematical analysis. To regard any but the simplest nuclei as conglomerations of different interacting particles treated individually, neutrons and protons, with their concomitant electrons, positrons, neutrinos, photons, and mesons—that was out of the question. The sheer complexity of the problem defeated such efforts. Some short cut was needed through the maze of complications if practical results were to be obtained to guide nuclear research. While the detailed studies must be pursued without letup, some simple, over-all princi-

ple was required lest the solution of pressing problems be delayed.

Here was an ideal setting for Bohr's unique genius. When
atomic theory falters, he helps it along with an admittedly
temporary theory which somehow proves dazzlingly successful. He did it in 1913 with his original atomic theory, and
again with his correspondence principle. He was now to
initiate another makeshift theory in 1936, his idea being
carried forward independently by the Russian theorist J.
Frenkel. Bohr is the great sustainer and tider-over of atomic
physics, a vital catalyst to keep the flickering mental flame
alive till it be self-sustaining. What theoretical physicist has
ever patched up and improvised so successfully and withal so
simply as Bohr? His earlier successes were no accidents. Here
was another seemingly jerry-built theory which was to prove
of phenomenal sturdiness.

What manner of thing was this new theory of Bohr and of
Frenkel? Was it some curious blend of the old and new,
some magic brew of quantum and classical ingredients to
parallel the atomic theory of 1913?

It was no brew of ill-matched essences, but a wholesome,
old-fashioned, purely classical theory. Yes. Classical, not
quantum.

If we were a little surprised at the recent mesonic signs of
second childhood, what shall we think of a theory that
actually likens the nucleus to a drop of water? What shall
we think of this final classical twist to our quantum tale?

At first one stands incredulous that such a classical, such
an utterly irrelevant model of the nucleus should have any
chance of success. But are not the nuclear forces largely
exchange forces, and are not the chemical binding forces

likewise exchange forces? Is it so startling, then, that there might be some analogy between the groupings of atoms and molecules and the groupings of fundamental particles in the nucleus? If water atoms and molecules cling together to form a drop, why should not nuclear particles likewise cling together to form a droplike nucleus? The known facts about nuclear forces show the analogy will be very close indeed. From a general, not too detailed point of view, the two would behave similarly, even such classical water-drop concepts as temperature, surface tension, and rippling waves being applicable to the nucleus. In the light of this analogy, familiar facts would take on new significance; thus particles would no longer be "shot out" from radioactive nuclei as of old but "evaporated" from them.

In 1939, with the world hovering ominously on the brink of war, there came the decisive experiment of O. Hahn and F. Strassmann in Germany which showed that barium resulted when neutrons bombarded uranium. Lise Meitner and O. R. Frisch, who fled Nazi Germany, found in the water-drop theory of the nucleus a picture already to hand to furnish the clue to what was taking place. With it they deciphered the cryptic message of the experiments. The uranium nucleus was undergoing fission, splitting violently apart, mass being converted into energy that was released in staggering quantities. When later it was found that neutrons too are released which could keep the fission process going spontaneously, physicists saw a new era opening up for mankind. For better or worse, nuclear energy of terrible potency was to be placed in the unready hands of man.

The fission idea of Meitner and Frisch was taken up by Bohr. He it was who foresaw the importance of the rare uranium isotope of mass number 235. With the American physicist J. A. Wheeler, he developed the water-drop model into a comprehensive mathematical theory of nuclear fission, even as Frenkel was doing so independently. This tentative, classical picture of the nucleus, proposed three years before, seemed almost to have been conceived with the new phenomena in mind, so extraordinarily apt did it prove. It alone could picture the process of nuclear fission. It alone could explain its mechanism and predict its various outcomes in this pressing hour of high urgency.

The picture it yields of the process of nuclear fission is one of extreme simplicity. When we say that the nucleus is like a drop of water we mean it not loosely and vaguely, as one would say it in conversation, but mathematically and precisely, having regard to its structure and internal stresses, for these are so very similar in the two cases that we may project the behavior of an electrically charged water drop into the behavior of a nucleus.

Nuclear fission was not an obvious concept. True, the presence of barium indicated that the bombarded uranium nucleus might have been split apart. But how could the almost negligible impact of a neutron have so cataclysmic an effect upon the nucleus? What terrible internal catastrophe could the gentle neutron have caused?

It was because they were able to imagine a plausible mechanism that Meitner and Frisch dared to suggest the possibility of fission. The forces in the nucleus are of two opposing kinds. On the one hand are electrical repulsions which, if unrestrained, would tear it violently apart. On the other are powerful attractive forces binding the nuclear particles one to another. These binding forces, however, have limitations. For example, unlike their antagonists the repulsions, they have extremely short reach. In a small nucleus they can easily hold the repulsions in check. But when the nucleus is large their reach proves inadequate and, what with other limitations, they cannot so readily dominate the forces of disruption which benefit greatly from the large electric charge of the heavier nucleus. A large nucleus, then, when likened to an electrified water drop, will correspond to one so big and so highly charged as to be on the verge of breaking apart. When an extra neutron is added to the nucleus it is as if a speck more water were added to the already swollen drop. The tendency to disrupt increases alarmingly. Let the neutron but be added with the gentle speed needed to set the nuclear drop aquivering and the chances are the drop will shake apart, forming two smaller nuclei which the victorious disruptive forces then cause to rush violently away, and, though Meitner and Frisch did not know it at the time, spattering forth a few small specks of nuclear matter—neutrons. For nuclear power slow neutrons are employed; in atomic bombs, fast ones. But the two pictures are similar in their essentials.

And here it was that the curtain fell, a curtain of dreary silence and suffocating secrecy hiding a deathly fear. What of the tremendous new theories which may have grown up in captivity, corralled behind the wire fences of Los Alamos or Oak Ridge? Such things are now military secrets, to be told by spies but not by scientists. Yet a corner of the curtain has been lifted to let some fragments of knowledge escape to the light, and despite the vast expenditure of effort and resources that went into the making of the atomic bomb, it has been said on high authority that no new theory has arisen therefrom to supplant or challenge the quantum.

Physicists are returning to the ways of peace. But the world, emerging from its ivory tower, now realizes that the innocent speculations of a Planck, an Einstein, or a Bohr may be charged with stupendous power for good or evil. The days of the nightmare are upon us, and science is in mortal peril of becoming an occult, unfertile priesthood, passing its mysteries on to chosen novitiates who meet stern tests and take the solemn vow of eternal silence. We can but hope the danger soon will pass, and someday, when the skies are brighter, science will again be free to stride forth boldly, in goodly fellowship, along its enchanted path into the unknown.

In truth, the story of the quantum is just begun. All that we have told is but a prologue. So many problems yet remain to be solved, so many questions are waiting to be answered. The picture is confused and tremendously exciting. It matters not that our theories are but temporary shelters from those icy winds of doubt and ignorance that chill the stoutest heart. Though they be destined to be forsaken by generations to come, they remain a wonderful adventure of the human mind, a wonderful exploration of the works of God. Crude and primitive though they may appear to men as yet unborn, they yet contain within themselves something of the eternal, and to our mortal gaze they stand a dazzling edifice of towering majesty, whose brilliance gladdens the soul and sends forth brave, struggling rays to pierce the murk and gloom that press around.

Here in such theories and discoveries is a revelation, all too scant, of the mighty wonder that is the universe. Here through the minds of our Einsteins and Bohrs we may dimly sense its structural beauty and cunning intricacy, its soaring poetry and its awe-inspiring grandeur and magnificence, with never a hint of its pain and tragic bestiality.

When at the empty dawn of all creation God created the primal essence energy, he endowed it with such subtle, miraculous potencies that, as from a seed that slowly comes to flower, there grew from it what we call space and time, and matter and radiation. Mightily, yet infinitesimally, there evolved a universe of coursing atoms and spacious nebulae. Energy coalesced into matter according to an immutable law so exquisitely contrived that amid the stupendous forces of writhing Nature there yet was found a gentle place and time, a small, quiet, friendly corner, to nurture fragile life. A trifling change in the laws of the universe, so small as to appear of negligible moment, and energy might have coalesced differently, might never even have coalesced at all. Deep down within the primal attributes of energy lay the rich promise of electrons and positrons, of protons, neutrons, mesons, and photons, of space and time and motion, of energy levels in nuclei and outside, of forces binding primary particles into atoms, atoms into molecules, and molecules into matter sustaining life and love and hate. What if the energy levels had been different? There might be no material universe. Nuclei now stable might be impossible structures. Oxygen, could it exist, might carry the deadly taint of radioactivity. Space and time might be cramped into narrow compass, with no vast regions of emptiness to protect the universe against its own explosive violence, with no vast aeons of time to let it slowly unfold and explore its innate heritage.

What little we understand of the deeper workings of the world is yet enough to reveal a sublime harmony beneath its turmoil and complexity. Our fragmentary knowledge is not lightly acquired. A meager handful of men is vouchsafed each generation with the precious gift of scientific insight, and we marvel at their powers. How much more, then, shall we marvel at the wondrous powers of God who created the heaven and the earth from a primal essence of such exquisite subtlety that with it he could fashion brains and minds afire with the divine gift of clairvoyance to penetrate his mysteries. If the mind of a mere Bohr or Einstein astounds us with its power, how may we begin to extol the glory of God who created them?

Alas, that fear and greed may pervert the incomparable blessing of nuclear energy. Alas, that so great a treasure should bring forth unparalleled crisis, and that in this moment of direst peril the sovereign nations of the world, archaic relics of a bygone age as remote in fact as it is near in time, should squabble over dangerous irrelevancies; dangerous irrelevancies of national sovereignty and individual power which, if not forever banished from the earth, will bring on us war of unthinkable horror and futility whose end will be utter destruction.

Almost overnight mankind can now plunge from the technological triumphs of an atomic age to the primitive barbarism of a desperate struggle for individual survival against the harsh forces of animal and inanimate nature. Already, from relatively minor causes, starvation and want spread darkly over the earth. Now is the terrible crisis of our civilization. Now is the fateful hour of high decision. For better or worse, We, the People of Earth, must choose our future. It can be fine and lovable, gentle and dignified, and filled with joy and wonder and thrilling discovery. Or it can be degraded and obscene, despairing and wretched beyond measure, with death and primitive misery stalking the land unchecked . . .

> I call heaven and earth to record this day against you, that I have set before you life and death, blessing and curs-
> .ing: therefore choose life, that both thou and thy seed may live.
>
> Deuteronomy 30:19

POSTSCRIPT 1959

"My grateful thanks go to A. Pais for clarifying various concepts for me, and to J. A. Wheeler for reading the postscript and making several helpful suggestions."

B. Hoffmann
Queens College
Flushing, N. Y.
June, 1958

POSTSCRIPT 1959

INTO a book that already contains a preface, a prologue, an inter-mezzo, and an epilogue, a postscript intrudes itself with ill grace. Some apology is needed for its presence.

Much has happened in the world at large during the ten years that have passed since this book first appeared, and much has happened, too, in the world of the quantum. The occasion of a reprinted edition gives me a chance to bring this book up to date, and if I do so by means of a postscript, it is with good reason. A postscript has valuable properties that a regular chapter lacks. For example, one can cram into it — indeed, one is expected to cram into it — not only the latest breathless news but also an assortment of items that one forgot to mention before. Above all, a postscript is informal; nobody expects it to be trim and orderly. Thus it is an ideal instrument for my purpose; for science is seldom tidy, except in retrospect. The events of the last ten years that touch on the quantum do not fall into a neat, inevitable pattern. They sprawl and flounder. They breach old boundaries in unexpected places that may or may not prove to lead to brave new worlds.

There is hazard in reporting scientific events so soon after they occur, while the stir and bustle they excite are still unsettled. The Nobel prize committee is well aware of this hazard, often delaying the award of a prize till a decade and more after the discovery for which it is awarded. Max Born, for example, did not receive his Nobel prize till 1954, almost thirty years after his interpretation of Schrödinger's smeared electron as a wave of probability. Delay is so far from being the exception that when T. D. Lee and C. N. Yang — of whom more later — were chosen for the Nobel prize a year after the appearance of their first paper on the possibility that parity might not be conserved, and within mere months of the spectacular experimental confirmation of their hypothesis, the event was hailed as extraordinary, albeit amply justified.

Which quantum events of the last ten years shall I tell about? Which shall I pass over lightly? Which shall I not mention at all? Whatever my decisions, time will surely mock them. For events have a knack of twisting unexpectedly.

The story of the Yukawa meson is an excellent case in point. Writing the epilogue in 1947, I told that in 1935 Yukawa had predicted the existence of a particle, the meson, and that shortly thereafter the cosmic ray experimenters had discovered that a meson actually existed. This seemed an eminently satisfactory situation (though I marred my account with remarks, now wholly inappropriate, about the resemblance of the meson equations to the equations of Maxwell). Yet even as I wrote, events were taking their unexpected turn. Three Italian cosmic ray experimenters, M. Conversi, E. Pancini, and O. Piccioni, made a disquieting discovery: the mesons they had been studying in the cosmic rays had little affinity for nuclear particles. This was a grievous blow to the theory of Yukawa, for mesons that had little affinity for nuclear particles could hardly be expected to bind such particles together. Thus the discovery of mesons in cosmic rays, which had once seemed so apt and speedy a confirmation of Yukawa's hypothesis, was now seen to have been no confirmation at all, but rather a case of mistaken identity. True, Yukawa had postulated a particle of intermediate mass, and a particle of intermediate mass had been observed. But the observed particle lacked the key property demanded by Yukawa: the ability to interact strongly with nuclear particles. The meson situation was once more untidy.

Yet even as these events were unfolding, another twist was developing. In those days, so soon after World War II, austerity was the rule in England. With money scarce, and cyclotrons and suchlike paraphernalia of nuclear research immensely costly, the physicist C. F. Powell employed a wonderfully inexpensive method of studying cosmic rays. Essentially, he merely left unopened packages of photographic plates lying around for several weeks — preferably on high mountains — developed them, and minutely analyzed the tracks left in them by the cosmic ray particles. In Powell's hands this technique gave remarkably detailed information and for his brilliantly parsimonious researches he was awarded the Nobel prize in 1950.

In 1947, Powell and two collaborators, the Brazilian C. M. G. Lattes and the Italian G. P. S. Occhialini, using this photographic technique, discovered mesons of a new kind in the cosmic rays in the high atmosphere. The new mesons, slightly heavier than the old, had extremely short lives; and those with positive charge, on decaying, gave birth to mesons of the old type. The new mesons, being primary, were called π (pi) mesons, and the secondary mesons to which they gave rise were called μ (mu) mesons.

One could perhaps deplore the added complication of now having two types of meson, but if nature was so constructed it was so constructed and one simply had to accept the fact. Had physicists been able to foresee later discoveries of yet other types of particles, including more mesons, they would have been less concerned about this slight increase in the complexity of the building material of the physical universe. Even so they had an immediate consolation, for the π mesons, unlike the old μ mesons, did interact strongly with nuclear particles. The π mesons could therefore be the mesons that Yukawa had postulated so many years before. Thus Yukawa's theory was vindicated after all, and Yukawa received the Nobel prize in 1949.

Yet this belated vindication in its turn proved dubious. The π meson failed to meet all the specifications originally laid down by Yukawa. For example, it apparently did not give birth to electrons*; because of this and related difficulties, and the puzzling existence of the μ meson, Yukawa's early hope of accounting for the emission of electrons from nuclei could not be fulfilled.

We shall have more to say about mesons later on. The theory of mesons has not trod a royal road. Yet despite difficulties and keen disappointments it has proved enormously fruitful in stimulating ideas and novel experiments. That it is still far from being capable of giving a satisfactory account of nuclear phenomena is less an indictment of the theory than a manifestation of the complexity of these phenomena. For no other nuclear theories are really satisfactory, though some have achieved excellent local successes.

For example, there is the idea of Bohr and Frenkel of treating the nucleus as though it were a drop of liquid composed of

---

* Time waits hardly at all to begin its mockery. Evidence has come from Geneva that the π meson does sometimes decay into an electron.

protons and neutrons — *nucleons*, as these particles are called. This idea and its offshoots are called *collective theories* because they treat the nucleus as a collective whole and are concerned with such overall effects as the vibrations in its shape and the circumstances under which these vibrations become so violent that the nucleus breaks into splatters.

Collective models of the nucleus have not lacked successes. But many details have eluded them. Take, for example, the seemingly haphazard numbers 2, 8, 20, 28, 50, 82, and 126. Physicists call them magic numbers, a name harking back to the days when the numbers were less well understood than they are now. The magic numbers have special nuclear significance; among the many hundreds of known nuclei, those containing just these numbers of neutrons or protons stand out from the rest because of their greater stability and other tell-tale signs. Clearly they reflect fundamental properties of the possible configurations of nuclear matter. They present a prime challenge to any nuclear theory. And one such theory has brilliantly met the challenge. This theory, the shell model, goes back to the earliest days of nuclear quantum theory. Its characteristic assumption is that each nucleon can be treated individually inside the nucleus, and that the motion of any one nucleon can be calculated by ignoring the individualities of all the other particles, lumping together the effects of all these other particles, and then, since these effects are not known in detail, replacing them by a sort of non-quantum spherical box or container within which the individual nuclear particle splashes around. This seemingly crude approximation had some initial successes; but they were not enough to overcome the awkwardness of its basic assumptions. Indeed, Bohr brought forward his liquid drop model in protest against the shell model, and the quick successes of the collective theories caused interest in the shell model to languish. In those days only a few of the lowest magic numbers were recognized. But in 1948 Maria Goeppert-Mayer, in America, collating known experimental results, showed impressively that the list was indeed magical and that it must include the further numbers 20, 50, 82, and 126. This had immediate repercussions. In Germany O. Haxel, J. H. D. Jensen, and H. E. Suess proposed a modification of the shell model of the nucleus, a modification proposed independently by Goeppert-Mayer. It was a curious modification: where, before, the effects of the other nucleons were replaced simply by a spherical container, now there was added a further effect of a magnetic sort, but one that could not be satisfactorily justified. Nevertheless the new effect could be so tailored to the needs of the moment that the theory was able to account splendidly not merely for the existence of magic numbers but for their actual numerical values and for the special nuclear properties that went with them. In 1950, J. Rainwater, in America, added further successes to the theory by distorting the spherical shape of the container.

In the shell model the protons and neutrons within the nucleus are treated much as the earlier quantum theorists treated the electrons surrounding the nucleus. The extranuclear electrons determine the chemical properties of atoms, and the earlier theorists, applying the Pauli exclusion principle to these electrons, were able to account for the exceptional chemical stability of the noble gases, helium, neon, argon, and the like, containing 2, 10, 18, 36, 54, and 86 electrons respectively. In essence, they accounted for magic numbers connected with the chemists' periodic table of the elements. Just so, the proponents of the shell model have been able to explain the

magic numbers connected with the physicists' more complicated table of the atomic nuclei.

The shell model has many considerable successes to its
credit — successes whose very impressiveness is a source of
puzzlement and embarrassment. For the theory seems obviously incorrect. The nucleus is so crowded with whirling
nucleons that one would expect their mutual jostlings to make
the smoothly regal nuclear motions contemplated by the shell
theorists quite unrealistic. Yet the successes of the theory are
there, and can not be laughed away. All is not dark, however.
Fermi made an illuminating suggestion: we know that the
Pauli exclusion principle prevents two nucleons from sharing
the same quantum state; with so many of the neighboring
states in the crowded nucleus already occupied, a jostled
nucleon has few convenient quantum states to go to, and so,
unless a jostle is unusually violent it will be unable to budge
the nucleon from its regal motion after all. While a quantum
state of motion is an elusive thing to visualize, the general idea
will be clear to anyone who has tried to budge the straphangers
in a crowded subway train during rush hour.

Shell models and collective models tend to complement
each other, each type working best where the other works
worst. Why not, then, try to fuse the best features of each
into a single theory — a *unified model* of the nucleus? The
idea was proposed in 1952 and proved singularly successful,
not only in explaining known effects but in predicting new
ones that were subsequently confirmed. Not the least pleasing
aspect of the unified model, and one that must have been
particularly gratifying to Niels Bohr, is that it was proposed
by his son Aage Bohr.

There are yet other nuclear models, modifications of the

basic types above. All are approximations, and each has its own niche of successes. But the overall picture, while rapidly becoming clearer and more detailed, is still one of probing rather than decisive penetration.

Nuclear forces are not properly understood. Even if they were, the crowded nucleus would pose formidable mathematical problems that would make exact calculations virtually impossible. This in itself might not be too serious, for most calculations in physics are perforce approximate ones, and in these days of giant electronic computers the drudgery of computation is no longer the deterrent it once was. But, though the nuclear models discussed above deal principally with protons and neutrons, there is still reason to believe that mesons are deeply involved in nuclear forces, and the numerical values of certain meson constants are so large that they cast grave doubts on the very validity of the customary approximation procedures and thus deprive the theorist of one of his most potent weapons. What profit is there in performing an intricate computation when the whole computational procedure is suspect right from the start? The surprising thing is that these computations sometimes give good results in spite of their inherent defects.

To add to the woes of the theorist, he has been overwhelmed by a veritable flood of new fundamental particles discovered in the last few years. They have mocked any hopes he may once have entertained that the structure of matter was on the verge of being clarified. Yet he has managed to keep bravely abreast of the flood, and has made heroic classifications and discovered tantalizing patterns among the new particles that hint of tremendous things as yet only dimly perceived. We shall tell of these later. Meanwhile there are other events to be reported.

Once upon a time, before the flood, in the far off days of 1928, when the world of the quantum was yet young and innocent, people believed that the material world was built of protons, electrons, and photons — just three types of particles; and in those days Dirac was trying to make an even simpler world — one built out of only two types of particles — by thinking of protons as absences of negative energy electrons in a monstrous ocean. This much we have already told, and we have told, too, how Dirac's absences of electrons proved to be not protons but positrons.

The idea of a monstrous ocean was awkward, to say the least; and the situation was not improved by the fact that, in principle, Dirac's equations referred to a single electron and ought not to encompass the teeming swarms needed for the monstrous ocean and its bubbly turmoil. Something needed to be done. But what?

Actually Dirac himself had already supplied the clue to the remedy. He had, as we know, enclosed Maxwellian light in a box so that it behaved like a collection of oscillators, and on these oscillators he had then clamped quantum properties. Where the Maxwellian theory of light had been a wave theory, the new quantized theory proved to be one that embraced both the wave and particle aspects and actually referred to hordes of photons.

The great success of Dirac's theory of light, its ingenuity, and its intrinsic air of rightness set men thinking. Heisenberg and Pauli, and Jordan and Wigner, and others began extending the idea and applying it to other types of waves.

What other types of waves? Well, there were the Schrö-dinger waves, for example, and the Dirac electron waves.

You will object, perhaps, that these are already quantum waves and represent particles as well as mere waves. In a sense this is so. Never mind, though. Just wait and see what happens. There were good reasons for wanting to modify them; so the theorists took the various types of quantum matter waves, among them the electron waves of Dirac, treated them as though they were "classical" waves on a par with the Maxwellian light waves, and by a manoeuvre analogous in principle to that used by Dirac on the light waves, clamped further quantum properties onto the matter waves — a process known as second quantization.

Perhaps highly successful quantum waves deserved to be spared the indignity of having this second quantization inflicted on them. But, as the results emphasized, the original Dirac electron waves, for example, really were on a quantum par with the Maxwellian light waves, and when subjected to second quantization they blossomed into vastly more potent entities capable of encompassing myriads of electrons and positrons in a turmoil of mutual annihilation and pair creation. Then, in 1934, Heisenberg showed how to dispense with the former monstrous ocean and its bubbles, and to treat electrons and positrons as co-equals in the second-quantized theory.

But one thing in particular marred the rejuvenated Dirac theory and cast a pall over even its most brilliant successes: when theorists tried to calculate certain quantities, for example the amount of energy associated with an electron, they obtained the ridiculous answer "infinity." Infinities infested the theory and insulted its creators. How, then, did the infinity-bedevilled theorists manage to perform calculations that could be tested against experimental data? With crossed fingers they introduced all sorts of special tricks and dodges to erode the infinities, or even snip them out. Their various ruses kept the infinities snarlingly at bay while the computations were being performed; but no one seriously believed in the validity of the ruses, and probably everyone who used them did so with an uneasy mathematical conscience. Nevertheless the computations could be made, and the results gave spectacular evidence of the aptness of the theory even as the infinities gave evidence of a deep-seated malady; deep-seated, and also pervasive, for though some of the infinities were inherited from the classical theories, others were the fruits of the second quantization itself; and both types could be blamed on the theory of relativity.

In 1943 Heisenberg, exasperated by the infinities, proposed a heroic new theory designed to avoid their very locale. At the age of 23, fresh from a stay in Copenhagen and deeply influenced by the profound, instinctive atomic wisdom of Bohr that has so surely moulded the world of the quantum, he had, as we know, renounced the unobservable electron orbits to build his matrix mechanics. Now he sought to repeat the triumph of his youth. Stay away, he urged, from the places where particles clash and infinities cluster. These places are dark and dangerous. They can not be directly observed. They are beclouded by uncertainties, and the happenings that we now imagine in them may be as fictitious as the old electron orbits. Stay with the tried and true, the things that are clearly seen and indubitable.

What are these things? Are there any at all? Yes. We can shoot particles at other particles and observe how they are affected by the collisions. This is, indeed, a principal mode of exploring atomic phenomena.

Back in 1937, Wheeler had introduced a mathematical quantity, the scattering matrix, or S-matrix, that was to prove ideally suited to Heisenberg's new purpose. Suppose we have a ψ that mathematically represents our beam of particles well before the collision, and another ψ that represents the scattered beam well after the collision. The happenings in the dangerous region are reflected in the changes in the beam, and these are reflected in the change from the first ψ to the second. Let us then fathom the dangerous region by studying the ψ change. Let us seek the rules that govern this change. First we must describe the change appropriately, and here is where the S-matrix comes in; for the S-matrix is an operator which, when applied to the first ψ, converts it into the second. Thus the S-matrix becomes the central object of the new theory, a veritable dossier of news from the danger zone.

What rules must the S-matrix obey? Heisenberg was loath to hunt for them among the old ideas and equations that were suspect in the dangerous regions. He sought to remain on safe ground by extracting self-sufficient rules for his S-matrix from various basic requirements: for example, that the rules must conform to demands of the theory of relativity, and that they should not imply that an effect can precede its cause.

It was an ambitious program, and a valiant one. But it did not succeed. General requirements such as those above failed to yield sufficiently detailed rules for a self-contained theory; indeed, the second requirement above proved singularly recalcitrant. Thus the infinities could not be by-passed in the manner Heisenberg had envisaged. But the S-matrix was to play a central role in later developments.

As for the infinities, they are still with us — but greatly tamed. Unable to avoid them, the theorists have learned to live with them, as we shall now tell.

When Dirac proposed his theory of the electron in 1928 he could not foresee the adventures that would befall it. Even without the later idea of the positron bubbles, it had extraordinary achievements to its credit. Did it not bring together quantum theory and relativity? Did it not account for the spin of the electron? Did it not yield an improvement on the Sommerfeld formula for the fine structure of the hydrogen spectrum — an improvement that was in excellent agreement with the intricate details of the observed spectrum?

Here was a theory that commanded respect. But not even the most impressively successful theories are immune to the disrespectful probings of the experimenters. How well did the Dirac theory really account for the fine structure of the hydrogen spectrum? Excellently well, apparently. The theory seemed to match the observations right up to the limits of precision of the measurements. But the finer details of the fine structure were not easy to observe, being blurred by the motions of the hydrogen atoms producing the spectra, and some observers in the late nineteen-thirties thought they detected discrepancies peeping through the slight fuzziness of the spectral lines. No one could be sure, for the measurements were difficult. One or two theorists made half-hearted attempts to explain the possible disagreement, but on the whole little attention was paid to the matter: life was so much simpler if one ignored unlikely discrepancies. Nevertheless a doubt had been raised.

In 1947 the American experimenter Willis E. Lamb, in collaboration with R. C. Retherford, put the question to a novel test. He used no prism or grating to produce a spectrum; he used no optical spectrum at all. Instead he studied the energy levels directly by a brilliantly ingenious method that largely avoided the blurring effect of the motions of the atoms and was capable of yielding results of extraordinary precision.

According to the Dirac theory of the electron, two of the lower energy levels of hydrogen should have exactly the same energy. Doubt had been raised as to whether the energies were in fact exactly equal. The Lamb-Retherford experiment was conclusive. It transformed what had formerly been a mere doubt into a glaring discrepancy. The celebrated Dirac theory of the electron was not in accord with the facts. Though the shift in the energy was small, the implications of its indubitable existence were profound. The phenomenon became known as the *Lamb shift*, and Lamb received the Nobel prize in 1955.

By a happy chance, a conference on the foundations of quantum mechanics was being held on Shelter Island in June of 1947. The startling news of the Lamb shift caused frenzied discussion among the participants; and out of the discussion came not only swift understanding but also a beautiful vindication.

First the cause of the Lamb shift was divined, and its locale in the theory seen to be the very heart of the danger zone, where it lay smothered by infinities. Next the Dutch physicist Kramers made a basic suggestion for dealing with the infinities. Then H. A. Bethe, who originally came from Alsace, quickly used it in a rough, non-relativistic calculation that gave highly satisfactory results. Upon which the American virtuoso Julian Schwinger, and others, began a series of more detailed, relativistic calculations that were to yield results in astoundingly close agreement not only with the experimental value of the Lamb shift but also with other anomalies into which we need not enter. And these dazzling results flowed from the very Dirac equations that had seemed to be placed under a pall of doubt.

Seldom has a crisis been so rapidly met and resolved. When the initial excitement died down it was found that one of Schwinger's basic mathematical ideas had been anticipated. The Japanese theorist S. Tomonaga had published it as far back as 1943, and its roots stretched back to earlier work by others. That the idea did not attract wide attention at the time may be attributed to its having been published in Japanese, yet the Swiss physicist Stueckelberg had actually discussed it as early as 1934; and Tomonaga had published an English version of his own paper in 1946; however the idea remained relatively unknown till Schwinger thought of it independently in the midst of crisis.

What was the secret of the Lamb shift, and how were the infinities dealt with? The secret lay in long-known phenomena associated with second quantization — phenomena in which particles are not permanent but may be created and destroyed.

We have all enjoyed stories about men on desert islands. And some of us have heard the old argument that it is impossible for a man to be on a desert island in the first place because his presence makes the island no longer desert. This verbal quibble applies, of course, to the world of the quantum too. For instance, we can not have an electron in a vacuum because the presence of the electron negates the vacuum. But in the quantum world we may not stop with this pleasantry. The situation is more serious, for the electron destroys the vacuum not merely by being present but by actually interacting with the vacuum and modifying it.

This is a strange idea. How can an electron interact with nothingness?

In the world of the quantum, particles are incessantly appearing and disappearing. What we would think of as empty space is a teeming, fluctuating nothingness, with photons appearing from nowhere and vanishing almost as soon as they are born, with electrons frothing up for brief moments from the monstrous ocean to create evanescent electron-positron pairs, and with sundry other particles adding to the confusion. Whence comes the energy for the creation of these particles? It is borrowed. And it is paid back before the default can be detected. That is why the lives of the particles must be extremely short. Were the particles to exist for an appreciable length of time the energy balance would be destroyed.

How does mere shortness of life save the energy balance? Through Heisenberg's indeterminacy principle. If the life is short, the time at which the particle exists is sharply defined. The more sharply we know the time, the less sharply we know the energy. And if the time is sufficiently sharp the energy becomes so fuzzy that the violations of the energy balance are completely obscured. There is a momentum balance to consider, too, but momenta are blurred when particles are precisely located, so this too presents no insurmountable obstacle to fleeting falsification of the books.

Evanescent, come-and-go particles that evade the laws of conservation of energy and momentum are called virtual particles. They were unthinkable in the pre-quantum era. Even in the quantum era they can exist only by virtue of the Heisenberg principle. But exist they do. And their role is far from negligible.

For example, if we place an electron in empty space it immediately conjures forth virtual photons, electrons, positrons, and other particles from the void. We have likened it to a man on a desert island. But we should liken it, rather, to a man on a picnic, beset by hordes of buzzing midges, gnats, and mosquitoes that seem to come from nowhere.

An electron in empty space does more than conjure forth quantum mosquitoes. It also disturbs the larvae: it affects the breeding ground, the monstrous ocean whence come the virtual electrons and positrons.

An electron clothed in a cloud of virtual particles would behave differently from one that was bare. The quantum theorists had been well aware of this, but they had shied away from it. For they well realized that the principal effect of the cloud of virtual particles would be to increase the effective mass of the electron, but when they calculated the increase it came out infinitely large. Since this infinite answer would smother everything, the theorists tended to ignore the secondary effects.

Under pressure of the Lamb shift, though, they looked more closely at these secondary effects — these gentle whispers drowned out by infinite cacophony — and realized that among them lay the most likely solution to the Lamb conundrum. In the old Dirac theory, it was as if, in marvelling at the energy of a man on a picnic fleeing from an enraged bull, one was so intent on the man's peril that one thought only of his sprinting feet and overlooked the mosquito-induced flailing of his swatting arms. Dirac's celebrated calculations of the energy levels of the hydrogen atom had been made for a bare electron moving under the electromagnetic influence of the nucleus in an old-fashioned, classical emptiness. But actually the electron was surrounded by virtual photons in a modern electromagnetic field teeming with quanta, and its interaction with this quantized field would involve not only the former items taken into account by Dirac but also new effects arising from the quantum aspects of the field. These last would shift the energy levels away from their former values, and so could perhaps account for the Lamb shift — as in fact it was found that they did.

To explore the possibility further the conferees had to devise a way to escape from the smothering dead weight of the infinities. Kramers took the decisive first step. He pointed out that we never measure the mass of a bare electron. The electron is always surrounded by its cloud of virtual particles and what we measure is always the total mass of the electron and its cloud. Our present theory is defective; it yields infinite mass for the cloud. Suppose we had a better theory that yielded a sensible mass. Then we would begin by assigning a mass to the bare electron, calculate the additional mass due to the cloud, and say that the combined mass was the mass we actually observe. Thus the bare mass with which we started — and, indeed, with which theories hitherto had both started and finished — was not the observed mass of the electron. The bare mass must be adjusted because of the cloud, a process called mass renormalization.

Now let us return from the delightful realms of fantasy where all problems are solved by a wonderful theory that does not yield infinities. Here we are back to earth, our theory imperfect and a Lamb shift urging us forward. We pick out the mathematical expression that corresponds to the mass of the cloud. It is infinite. True. But that is not our fault. It is the fault of the faulty theory. In a better theory this expression would have a suitable value. So let us take this infinite quantity, add it to the bare mass of the electron, and simply say dogmatically that we will call the result the observed mass of the electron. Then all we need do is to replace this infinite mathematical quantity by a quite small number furnished us by the experimenters. It is cheating, of course. But inspired cheating. And it works miracles. For now we can keep the infinities at bay and calculate the delicate secondary effects hitherto smothered by them. Bethe's calculations were possible only because of this trick. And all the subsequent renormalization calculations were utterly dependent on it too.

Renormalization is not confined to mass; it applies to electric charge too, for example. Though it has an aspect of chicanery when applied to infinities, it is far from being a mere trick. Even if we had a theory giving finite answers we should still have to use the renormalization technique — but in that case we could use it with full mathematical propriety.

With the aid of the renormalization technique, the quantum theory of electrons, positrons, and photons was brought to so high a degree of perfection that no single phenomenon within its compass remained unaccounted for. Its predictions were confirmed down to the finest observable detail by experiments of the utmost refinement. The Dirac theory of the electron had indeed come into its own, forming in conjunction with the quantized theory of the electromagnetic field a triumphant theory able to meet all experimental challenges with a dazzling precision that placed it among the most successful physical theories of all time.

And yet the infinities are still there, lurking and snarling, tamed but unvanquished. Renormalization pushes the accomplishments of the theory beyond the limitations of the theory

itself, and shows that deep down there is something essentially right beneath the morass of infinities that are certainly
wrong. It does not show us how to prevent the infinities
from occurring, but only how to live with them. It does not
point the way to a new theory free of the curse of the infinities
— or if it does, no one has yet interpreted its hints correctly.
Rather it seems to take us farther than ever into a cul-de-sac.
The audacious juggling with infinities is extraordinarily brilliant. But its brilliance seems to illuminate a blind alley.

There is another major theoretical advance to report. It
begins in the year 1948, though its roots go farther back; and
it is nothing less than a new mathematical formulation of
quantum mechanics — by an American youngster, Richard P.
Feynman, then barely out of graduate school. It has to do in
part with the notion of time.

We really know very little about time, yet it is the stuff
of which life is made. We live our lives trapped in the fleeting
present, a timeless boundary between past and future whose
flow is the very essence of time. Memory may cheat the present
by bringing fond recollections; but we can not return to the
days of our youth, for time hastens on whether we will it or
not. Nor can we speed time in its flight; we must wait
patiently till tomorrow comes.

Yet in imagination we can escape the grip of the present
and roam freely in the realm of time. We can move forward
and back — peer into future and past — and map the domain
as if it were ever present. To restore the semblance of the
flow of time we view our static map with an eye that roves
from past to future, much as a doctor might scan the fever
chart of a patient.

The mapping of time is a commonplace in science, and

furnishes valuable insights. We do not always use it, though. For instance, when, as is customary in quantum theory, we think of particles as progressive waves we think of them as in the perennial, flowing present, and write equations expressing the manner in which they evolve as they go from one present to a present a moment later.

Young Feynman let go the apron strings of the present by making a map of the past and the future and treating the particles therein as particles rather than waves. By so doing he forsook the firmly entrenched tradition of Hamilton with its emphasis on progressive waves, and returned, in essence, to the earlier way of Lagrange.

In everyday life when we make maps of time, for example in graphs of temperature and rainfall, and of business conditions and the like, we usually think of time as progressing from left to right. Feynman, schooled in the theory of relativity, followed the relativistic custom of representing time as flowing from bottom to top of his map. In theory, his map had four dimensions, three belonging to space and one to time; but for practical purposes he sketched maps with two dimensions only, one of space and one of time.

Let us take a blank sheet of graph paper and represent space horizontally and time vertically. Suppose we place a dot somewhere on the graph paper. What sort of thing will it represent? Remember that time flows upwards. To bring this vividly to mind let us view the graph paper through a "time slot," a narrow horizontal slot that moves steadily upwards, from bottom to top of the paper. What we see through the slot at any instant is space in the present; when looking through the slot we are blind to both past and future. With only a single dot on the graph paper we shall most of the time see nothing at all through our time slot. Then the dot will suddenly appear for one brief moment and be gone. Clearly the dot will not adequately represent a particle. A particle persists in time. To represent it on our map we must draw a line — a *world line*, in relativistic parlance. Suppose we draw a vertical line. Then as we view it through our travelling time slot we see a persisting dot that remains at rest in the slot. Thus our vertical world line can represent a particle at rest. Do we want a particle that moves to the right? Then we draw a world line that slopes to the right, for when we view it through our time slot we see a dot moving to the right. For a particle that moves to the left we draw a world line sloping to the left. A zig-zag line would represent a particle that changed speed abruptly, as though colliding with other objects.

What of a pair of particles suddenly created? A V-shaped mark can take care of that. For as we look through our moving time slot we at first see nothing; then when we reach the apex of the V we see a dot that at once becomes two dots scurrying away from each other in opposite directions. For two particles that annihilate each other we need an inverted V.

Now that we understand the rules, let us try our hands with a little story. Here is the scenario:

We start with a lone electron, E_1. Suddenly a positron-electron pair, P_2, E_2, is created, and P_2 and E_2 speed apart. Now we have three particles. But after a while positron P_2 encounters the original electron E_1 and the two annihilate each other. This leaves the second electron, E_2, as the sole survivor. In real life photons would be involved along with the electrons and positrons, but we shall concentrate only on E_1, E_2, and P_2 here.

How will our space-time map look? Here is a possible version:

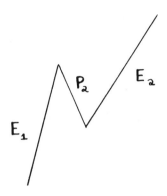

View this through the travelling time slot and you will easily see that it follows the scenario faithfully.

So far there is nothing at all new. Three world lines to represent three particles — that is standard practice.

But in his graduate student days, Feynman had collaborated with Wheeler on a theory wherein certain effects could precede their causes.* And during their collaboration Wheeler had had a remarkable idea. Like the boy who cried "The Emperor has no clothes," he had suddenly realized there was only one world line, not three. Look at it. One zig-zag line. Count it.

No doubt we retort in horror that this makes the P_2 part run backwards in time; that the graph has three sensible world lines like this

*For reasons connected with the war, part III of their study appeared four years before part II. Could one reasonably ask for a more telling experimental confirmation of their thesis of topsy-turvy time?

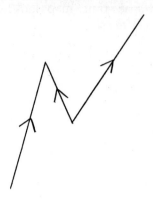

not one nonsensical one like this

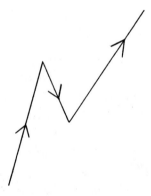

But let us not dismiss the idea so hastily. Admittedly the new scenario is vastly different from the original one, running, as it does, as follows:

We start with a lone electron, E_1. At a particular moment the electron undergoes a collision so violent that it causes the electron to speed into the past. Then — if that is the word — at an earlier moment it undergoes another — or perhaps we should say a previous — violent collision that causes it to go forward into the future.

Yet when viewed through our time slot the new diagram tells the same story as the old one did.

Note how one zig-zag world line can give rise to the simultaneous presence of three particles. When Wheeler first had his idea he saw in a flash a stupendous cosmic pattern: a single electron shuttling back and forth, back and forth, back and forth on the loom of time to weave a rich tapestry containing perhaps all the electrons and positrons in the world.

Given the outlandish idea that an electron might travel towards the past, we can easily see how the electron when so moving would appear as a positron moving towards the future. For an electromagnetic field that pushes an electron in a particular direction pushes a positron, which has opposite electric charge, in the opposite direction; and a particle moving to the left as time progresses moves to the right when we run time backwards.

Feynman did dazzling things with the zig-zag world lines. He showed, for instance, that there is an essential similarity between

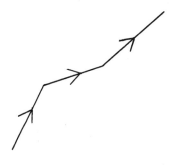

and

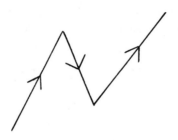

even though the physical processes seem at first glance quite dissimilar. In each case an electron undergoes two collisions, with objects not represented in the diagram. In the former the world line points throughout towards the future, and when we look at it through our moving time slot we see the electron make two abrupt changes of speed. In the latter the collisions, being more violent, reverse the direction in time, and when we look through our time slot we see pair creation and annihilation. In a sense the only difference between the two cases is that in the former the collisions occur in the expected order while in the latter the second collision occurs before the first. Does that phrase "the only difference" strike you as an understatement? It is, of course. Yet not as great a one as it seems. For Feynman did more than just point out that the diagrams are similar — zig-zags with two bends each. He demonstrated that the mathematics of the mild collisions was in principle the same as that of the violent ones, so that the two types could indeed be treated on an equal footing.

And he went much further than this. He associated probabilities with the various possible world lines and developed thereby a complete reformulation of quantum mechanics.

The older formulations reflected our entrapment in the present by viewing the world through a moving time slot and describing the evolution of waves from moment to moment; but Feynman took an Olympian view of time. Where the earlier formulations groped their way into the future as though through a fog, Feynman strode boldly, basing his strategy on time maps and following individual particles through their zigs and zags as readily towards the past as towards the future. In his world lines for photons, electrons, mesons, and the like, he had a graphic accounting system for the hitherto confusing tangle of collisions, creations, and annihilations, both real and virtual, that might occur. With each possible idiosyncrasy of the time map he was able to associate a corresponding mathematical expression, so that his time maps became simple campaign outlines for complex mathematical investigations.

Previously in solving their quantum mechanical equations, the theorists had often resorted to mathematical manoeuvres that could not readily be interpreted in physical terms. But Feynman's graphs described the actual physical happenings, and his mathematics, paralleling the convolutions of his graphs, remained intimately related to the physical processes under discussion, each mathematical term having its direct physical counterpart. Where the earlier calculations became lost in the terrain so confusingly glimpsed through the moving time slot, Feynman, with his overall picture, could steer his mathematics through hitherto impenetrable complications. The Feynman graphs, and the mathematical techniques based on them, have become invaluable basic tools of modern quantum theory, transforming its outlook and vastly increasing its power.

In his theory, Feynman was concerned with the life histories of particles and followed them through their various collisions. Thus he related their behavior before the collisions to their behavior after the collisions, and in this sense he had an S-matrix theory. But it differed from the theory envisaged by Heisenberg. It did not avoid the danger regions. On the contrary, it followed the particles wherever they went and took its chances with the infinities. Again, Heisenberg had sought to create a new physical theory free from some of the pressing defects of the old. But Feynman's theory, for all its remarkable novelty, was proved to be basically equivalent to the old, the proof of this equivalence being due principally to F. J. Dyson, a youthful theorist who came from England to work in America and fell under Feynman's influence just at the crucial time when the new formulation was being developed.

It is time now to tell briefly about the new particles that have been discovered in such unexpected abundance. To follow the chronological order of their discovery would be unnecessarily confusing. We shall group them a little to make their pattern clearer.

First, then, let us mention the experimental detection of the elusive neutrino. With the advent of atomic piles, neutrinos came to be in plentiful supply; for in a typical large pile the power leakage due to neutrinos is enough to light a small town. One might imagine, therefore, that the supply was ample for the experimenters bent on detecting the neutrinos. But the ability of neutrinos to penetrate the massive shielding surrounding a pile is striking evidence of the difficulty of stopping neutrinos in their headlong flight to the ends of the universe. So minute a proportion of the fleeing neutrinos could be entrapped that the supply was barely enough for the experimenters. It required an experiment of heroic proportions, culminating three years of work by F. Reines and C. L. Cowan, and others, before the neutrino was finally detected at Los Alamos in 1956.

The neutrino, of course, was an expected particle. Yet it had eluded the experimenters for more than twenty years after its existence was first suspected. There was another particle that eluded the experimenters even longer. The theorists had found that their equations possessed a certain symmetry between positive and negative charge known as invariance under charge conjugation. This told them that to every type of particle there ought to correspond an anti-particle. A positron, as we already know, is an anti-electron; and this means that the electron automatically qualifies as an anti-positron. But the invariance under charge conjugation implied that all particles, including even electrically neutral ones, should have their anti-particles. In Feynman graphs, for example, particles and anti-particles are represented by world lines pointing in opposite time directions, and when a particle and its anti-particle meet they annihilate each other with enormous release of energy.

In particular, since protons exist, there ought to be anti-protons too. An anti-proton would have the same mass as a proton but an opposite electric charge. No such particle was known. The experimenters had searched diligently for anti-protons, but for years they had searched in vain, and doubts arose as to whether the anti-proton existed. With the passage of time, though, the synchrocyclotrons and other atom-smashing machines grew in power till at last they were capable of producing the enormous energies needed to create proton and

anti-proton pairs — if anti-protons existed, that is. Indeed, the
Bevatron at Berkeley, California, was designed with anti-
protons in mind. And in 1955 a team of Berkeley scientists,
headed by E. Segré, succeeded, in a brilliantly conceived experiment, in producing protons and anti-protons and, what
was crucial, in identifying the latter unmistakably as anti-
protons. The achievement was greeted with admiration min-
gled with relief; relief because the symmetry of the theory
was now confirmed. Anti-neutrons quickly followed.

Yet all was not symmetric. Anti-protons have negligible
chance of survival. They are quickly annihilated by ordinary
protons, the latter being in ample supply. If there is complete
symmetry between protons and anti-protons, why are protons
so common and anti-protons so rare? A possible answer is
simply that the universe always has had far more protons than
anti-protons, so that after a possible initial holocaust of mutual
annihilations protons have survived in large numbers and anti-
protons hardly at all. This is a lop-sided answer. A more sym-
metrical one is that the numbers of protons and anti-protons
are comparable, but that we happen to be in a part of the
universe where the protons predominate. Predominance of
protons carries with it such effects as predominance of electrons over positrons. So if the second answer is correct we
must picture the universe as containing regions occupied by
matter anti to ours. A conjunction of two such dissimilar
regions would be catastrophic.

Up to about 1950, although their number and diversity
bloated the appearance of atomic physics, almost all of the
known fundamental particles could be encompassed within
accepted ideas; only the μ mesons seemed misfits. But then
came a deluge of unexpected and unwelcome new particles

that the theorists, with despairing candor, quickly called "strange particles." Some were heavier than protons. The others, called K particles, were new types of mesons, heavier than the old mesons but lighter than protons. The immediate strangeness of these new particles lay in their lifetimes, which were in the neighborhood of a billionth of a second. Short though this may seem, it was, according to accepted ideas, actually far too long — tens, even hundreds of billions of times too long. It was, indeed, a lifetime!

The long lifetimes had disconcerting implications. Gravitational and electromagnetic forces have long been known, the former being intrinsically many billions of billions of billions of billions of times weaker than the latter. In the quantum era other types of forces were discovered, and after a while people began to notice that they fell into two widely disparate groups, being either very strong or very weak, with none in between. The strong forces, associated with processes involving nucleons and others of the heavier particles were intrinsically more than a hundred times stronger than the electromagnetic forces, being indeed the strongest forces of which we have any knowledge; the weak forces, intrinsically about a hundred thousand billion times weaker than the strong forces, were originally recognized as being associated with processes involving neutrinos.

Just as a stiff spring vibrates faster than a pliant one, so do processes involving strong interactions go intrinsically faster than those involving weak ones. Thus particles whose decays are governed by strong interactions should have much shorter lifetimes than particles that decay by weak ones. Now the problem of the strange particles lay in this: since they are produced in strong interactions, they are certainly susceptible to the influence of strong interactions, and should therefore decay by strong interactions; but they have lifetimes characteristic of weak interactions. And in addition to the numerical discrepancy of the lifetimes, there was a qualitative discrepancy, for there seemed no room for conventional weak interactions anyway since no neutrinos were involved in the decay. The strange particles were gawky intruders that simply did not fit into the accepted scheme of things.

In 1952, at the Institute for Advanced Study, the theorist A. Pais, who came to America from Holland, saw a way out of the difficulty. Fresh from the stimulus of an international conference at Rochester, N. Y. largely devoted to the matter, and benefiting from elaborate studies that had been made of the strange particles, particularly by Y. Namba, S. Ōneda, and other Japanese physicists, Pais suggested that there must be some rule — some law of nature — that prevents strange particles from being produced singly. If, for example, two different types of strange particles were produced in a strong interaction, and the two particles immediately moved away from each other, they could slip through the fingers of the life-shortening, two-particles-at-a-time grip of the strong interaction and live to a ripe old age of a billionth of a second. Their leisurely decline would then have to involve a new type of weak interaction to be added to the neutrino group.

This idea had important repercussions. It resolved the paradox of the lifetimes; it revealed the existence of a new group of weak interactions not necessarily involving neutrinos; and, above all, it posed a formidable theoretical problem: why could strange particles not be formed singly in strong interactions?

The promising beginning of an answer came a year later,

when the American M. Gell-Mann and the Japanese scientist K. Nishijima independently proposed a scheme for bringing a measure of order to the crowd of fundamental particles. It introduced a new quantum number, S, measuring, in whole numbers, a baffling quantity that is aptly called the *strangeness*. What it meant physically was obscure — and still is. But it beautifully clarified the pattern of the fundamental particles, and even required the existence of further particles that were later detected experimentally. Thus the price of order was an increase in the number of particles; but this was really no price at all, for the new particles would anyway have been detected sooner or later.

Strangeness may seem to belong more to poetry than physics. But do not be misled by a word. Whatever it may be, strangeness is lumpy stuff: so far, it has been found only in units of -2, -1, 0, 1, and 2. It is fairly durable stuff, too, for it can be neither created nor destroyed in strong interactions, though, curiously, this is not the case in weak interactions.

Let us confine ourselves to strong interactions for a moment. Of the particles that interact strongly, the nucleons and π mesons have S equal to zero; this being the badge of their unstrangeness; the rest are all strange in varying degrees. Suppose two unstrange particles collide. There is zero strangeness initially. Therefore there must be zero strangeness after the collision. But we can not have zero strangeness with only one strange particle; we need at least two — for example, one with $S = 1$ and another with $S = -1$. Thus the rule discerned a year before by Pais was seen to be the law of conservation of strangeness in strong interactions. This law was not invented *ad hoc* to satisfy Pais' requirement. It was a consequence of quite different things having to do with electrical forces. Nor is this by any means the only instance of exquisite dovetailing in the emerging pattern of the fundamental particles. The pattern holds together well. It is not jerry-built.

Yet there are far too many types of fundamental particles for comfort. How many? There comes a time when the truth can be embarrassing. Let me answer with the classic feminine phrase "over twenty-one." Thus do I keep my estimate fresh for many a decade and protect myself against the flow of new discoveries threatened when giant accelerators now abuilding are completed.

My protection is only partial though. I am still vulnerable in a way no woman is. For surely there is some deeper unity beneath the present multiplicity, and should this be discovered it could reduce the number of really fundamental particles well below twenty-one. Even without such a principle the number of particles has not always risen, as the following, final item in this postscript will now show.

Imagine an experimenter who, after spending long months setting up his apparatus, is about to perform his experiment to test a law of physics. On his way to the laboratory he has an attack of appendicitis and is rushed to the hospital. Two weeks later he performs his experiment.

In writing up his account of the experiment for publication, he does not mention his appendectomy and the delay it caused. Why not?

Because it is irrelevant. Obviously.

Obviously, indeed. Yet there is an important principle behind this "obviously": the precise moment at which one makes an experiment to test a physical law is usually irrelevant so far as the testing of that law is concerned.

Whether the experiment is performed now or a couple of

weeks hence, makes no essential difference in the results obtained; the laws of physics will not change in the meantime. At least we hope not; and more than just hope, we assume not. If the experiment is an important one, other physicists, at different places and different times, will repeat it, making allowances for any accidental items like differences of temperature and atmospheric pressure; and, if the original experiment was an honest one, they will get the same essential results. Even if the original experimenter repeats his own experiment in his own laboratory he does so not only at a different time but also at a vastly different place; for the earth does not stand still in the heavens.

The irrelevance of location in time and space will be reflected in the mathematical form of the equations that express the basic laws of physics; the equations will have certain simple characteristics that ensure that the laws they express are the same at any time or place.

So far we seem to be saying nothing particularly exciting. But now comes a profound mathematical consequence. If our equations reflect the irrelevance of location in time, then, automatically, they imply a law of conservation — and the conservation turns out to be a conservation of energy. If our equations reflect irrelevance of location in space, there must be a corresponding law of conservation, in this case a conservation of momentum. Conservation laws are linked to irrelevancies.

That position and momentum should prove to be partners here, or that time and energy should, ought not to surprise us unduly. We have already met them as partners in indeterminacy; and their partnerships go back to the classical mechanics of the pre-quantum era.

There are other irrelevancies, and other conservation laws.

For example, it does not matter fundamentally which way round we place our apparatus; experimenters in America, Australia, England, and Russia can check each others' results though the directions of their ups and downs are quite different, as are those of their easts and wests, and their norths and souths. Coupled with this irrelevance of orientation in space is the law of conservation of angular momentum.

Again, we know from experiment that electric charge is conserved; a positron, for example, can annihilate an electron, but the total charge is not thereby altered. Corresponding to this conservation there is an irrelevance of a curious sort. It is called independence of gauge, or gauge invariance, and can be thought of as an irrelevance of location on a hypothetical mathematical line known as the gauge space. This one-dimensional gauge space is not akin to ordinary space. Yet a unified field theory of gravitation and electromagnetism has been constructed in relativity by considering it as an additional dimension making, with space and time, a five-dimensional world. Thus though the gauge space may seem like something of a mathematical fiction, it is, like many other such fictions, a powerful one. It should be, too, being related to so fundamental a law as the conservation of electric charge. Moreover deeper conservations suggest that two hypothetical dimensions may be involved, forming an entity known as isotopic spin space.

There is one irrelevance that is particularly relevant here: the irrelevance of handedness. If, instead of viewing nature directly, we view it in a mirror, thereby interchanging right-handedness and left-handedness, we can expect to notice no difference in the basic physical laws. This irrelevance differs from the previous ones in not leading to a conservation law in classical theory. It does lead to one in quantum theory, though, as E. Wigner showed as far back as 1927; a curious one called the conservation of parity.

Parity is not a readily visualizable entity. It is a quality of evenness or oddness. Particles in various states have either even or odd parity, and the rules for combining parities happen to be the same as those for combining even and odd numbers. For example, two even numbers add up to an even number; correspondingly, a system consisting of two smaller systems, each of even parity, has even parity. Again, two odd numbers add up to an even number; and, correspondingly, two systems of odd parity form a system having even parity. Further, an odd and even number add up to an odd number; and a system of odd parity and one of even parity yield a system of odd parity.

Suppose a particle of even parity disintegrates into two particles. Then the law of conservation of parity tells us that these two particles must form a system of even parity; that is, the two particles must both be of even parity or else both be of odd parity. But a particle of odd parity breaking up into two particles must yield one particle of even parity and one of odd parity; otherwise the total parity would be altered.

Thus the law of conservation of parity limits the number of possibilities, and it has proved invaluable in quantum physics, showing why many processes were never observed, their non-occurrence being otherwise incomprehensible. It was first noticed by the theorist Otto Laporte in Germany back in 1924, in connection with an analysis of the extremely complex spectrum of iron. This was in the era of the Bohr theory, before the explosive emergence of the new quantum theory, and the principle has had an extensive and honorable career since then.

Without it, for example, the complex disintegration scheme of the fundamental particles would have seemed wildly capricious.

Among the strange particles were some called θ (theta) particles and others called τ (tau) particles. They were K mesons.

Now a charged theta could decay into two π mesons; a charged tau into three. Since the parities of the π mesons are odd, the law of conservation of parity showed that the θ's should have even parity and the τ's odd.

But this conflicted with other information about the θ's and τ's that was coming in from the experimenters. The masses of θ's and τ's, for example, were about the same, and so were their lifetimes. With much cogent evidence suggesting that the θ's and τ's were really the same types of particles, the pronouncement of parity that they were different was unpleasantly jarring to physicists. The anomaly was discovered in England in 1953 by the theorist R. Dalitz in a masterly analysis of the available experimental evidence, and as better measurements were made they served only to accentuate the conflict. Soon the θ-τ puzzle had become one of the central scientific issues of the day. It was a prime topic of discussion at the 1956 International Conference at Rochester, N. Y., and with no satisfactory solution forthcoming, some of the participants even wondered whether the law of conservation of parity might be false. This was a striking measure of their desperation, for the evidence for the validity of the parity law was particularly strong.

Stimulated by the discussions at the conference, two young Chinese theorists, Tsung Dao Lee, of Columbia University, and Chen Ning Yang, of the Institute for Advanced Study, decided to examine together, with skeptical eyes, the seemingly overwhelming evidence in favor of parity conservation.

The evidence was indeed strong; yet weakness was to be its undoing. For, on meticulously assessing the extensive experimental data, Lee and Yang made a staggering discovery: though the evidence for the conservation of parity was compelling in the realm of strong interactions and electromagnetic interactions, in the realm of the weak interactions it was inconclusive.

Now the center of the θ-τ puzzle lay among weak interactions. Thus there was here a hint of a possibility of a tiny loophole. This was enough for Lee and Yang. Boldly, though with understandable signs of trepidation, for they were staking their reputations on a possibly ludicrous gamble, they took two enormous steps: first they suggested that parity might not be conserved in any of the weak interactions; and then they brought this speculation down to earth by showing specifically how it could be tested by experiment.

Nothing could better indicate the audacity of Lee and Yang than the fact that their proposals were promptly pooh-poohed by Pauli.

Lee and Yang were theorists. They could point out where to look for possible non-conservation of parity, and what particular symptoms to look for. And in principle their suggestion was simple enough: test whether processes involving weak interactions have definite handedness; see, in fact, whether the mirror images of such processes are physically impossible. But designing feasible experiments and performing them called for other talents.

So they took their problem to the experimenters, notably their gifted Chinese colleague C. S. Wu at Columbia University. With her mastery of the resources of the experimental art she designed a feasible experiment. First, atoms of cobalt 60 were to be lined up with their spins parallel — a difficult two-stage operation calling for an ingenious use of interactions with other atoms. Then, to reduce the trembling that is heat, everything was, so to speak, to be frozen in place for the few minutes during which the crucial measurements were to be made. Therefore E. Ambler, the American cryogenics expert at the Bureau of Standards in Washington, D. C., was called in, and the experiment was transferred to his laboratory where low temperature facilities were available. There Wu and Ambler and their coworkers girdled the heart of their experiment with an electric current. The current produced a magnetic field; the magnetic field lined up atoms in crystals of cerium magnesium nitrate; and these atoms in turn lined up cobalt atoms that had been incorporated into the crystal surfaces. The experimenters froze the lined-up atoms into relative quiescence. And they watched the directions in which electrons came off as the cobalt underwent radioactive decay — a seemingly trifling matter, yet a fateful one.

Place your watch on the table face up, and imagine that the electric current aligning the cobalt atoms flows clockwise around the rim of the watch. View this in a vertical mirror and the current will seem to flow in the opposite direction. Suppose the electrons come off equally upwards and downwards. Then the mirror image will be equivalent to the actuality, being just the actuality upside down. But if more electrons come off upwards than downwards, or vice versa, the actuality will have a definite handedness and its mirror image will differ from it as a right hand does from a left. Wu and Ambler and their co–workers found that the electrons did not come off equally

upwards and downwards. The actuality was intrinsically different from its mirror image. Lee and Yang were vindicated.
Handedness was important. Parity was not conserved. And
this was in a weak interaction, each electron given off being
accompanied by a neutrino.

Even as this experiment was in progress a quite different
one was being performed using the Columbia cyclotron. This
too corroborated the daring conjecture of Lee and Yang, the
results of the two experiments actually being published
simultaneously.

Lee and Yang were vindicated. Yet, ironically, the experiments that brought them fame did not directly resolve the
θ-τ paradox that had inspired them. True, the experiments
were concerned with weak interactions. But, for practical reasons, they dealt with weak interactions involving neutrinos.
They could thus be interpreted as showing that neutrinos
possess definite handedness. But no neutrinos were involved
in the θ-τ puzzle, the weak interactions there being of a non-
neutrino kind.

Once the parity rampart had been breached, however, evidence began to pour in from all sides confirming the violation
of parity conservation in weak interactions. And soon direct
evidence was found that handedness is significant in certain
processes involving weak interactions of the non-neutrino
sort, though still not the actual θ and τ interactions. Thus we
now know from direct experimental evidence that parity non-
conservation is not confined to neutrino processes. And in the
face of so much evidence, the θ-τ puzzle ceases to be a puzzle
and becomes rather the first of a rapidly increasing list of
experimental confirmations of the non-conservation of parity
in weak interactions of all types.

Evidently, then, all weak interactions are kin. But they seem to have disparate origins, and why they should be kin is a mystery to which at present there seems to be not even an inkling of a solution.

With parity conservation gone, handedness appears dominant and physics seems destined to denounce its mirror image as unphysical.

But softly. Not so fast. Handedness does not yet have the upper hand. There still may be a sort of mirror symmetry in the world. The laws of physics may still be their own mirror image, as Lee and Yang, and, independently, the Russian theorist L. Landau pointed out. Let us give our mirror magical power. When it interchanges right and left, let it also interchange matter and anti-matter. Then — unless our basic theory is playing us false — physics viewed in the magic mirror can still be valid physics; the laws governing left-handed matter, for example, can be the same as those governing right-handed anti-matter despite the non-conservation of parity.

Parity non-conservation has thrown physics into a turmoil that is still raging, but we may not tell here of the many further ramifications of the mirror symmetry problem. This postscript is already far too long, even though much has been omitted that could make excellent claim to be included. For example, there is nothing here about the various "non-local" theories of Yukawa and others. Nor about the recent attempt of Heisenberg to formulate a simple, all-embracing theory to account for the existence and properties of the fundamental particles. Nor have I mentioned recent attempts to reinstate causality among the basic principles of quantum physics, notably by D. Bohm, that have inspired de Broglie to extend his early ideas of a causal interpretation of quantum theory —

fascinating matters, but their outcome remains in doubt. Isotopic spin, which is related to strangeness, has received only niggardly attention, though it seems destined to play a significant role in future developments. And for all that this postscript may seem to be packed full to overflowing, it merely skims a few of the high spots of a chaotically surging tale.

But now my space has all run out and I must bring this postscript to an end.

Adieu.

INDEX

CATALOGUE OF DOVER BOOKS

(*) The more difficult books are indicated by an asterisk

Books Explaining Science and Mathematics

WHAT IS SCIENCE?, N. Campbell. The role of experiment and measurement, the function of mathematics, the nature of scientific laws, the difference between laws and theories, the limitations of science, and many similarly provocative topics are treated clearly and without technicalities by an eminent scientist. "Still an excellent introduction to scientific philosophy," H. Margenau in PHYSICS TODAY. "A first-rate primer . . . deserves a wide audience," SCIENTIFIC AMERICAN. 192pp. 5⅜ x 8.　　S43 Paperbound **$1.25**

THE NATURE OF PHYSICAL THEORY, P. W. Bridgman. A Nobel Laureate's clear, non-technical lectures on difficulties and paradoxes connected with frontier research on the physical sciences. Concerned with such central concepts as thought, logic, mathematics, relativity, probability, wave mechanics, etc. he analyzes the contributions of such men as Newton, Einstein, Bohr, Heisenberg, and many others. "Lucid and entertaining . . . recommended to anyone who wants to get some insight into current philosophies of science," THE NEW PHILOSOPHY. Index. xi + 138pp. 5⅜ x 8.　　S33 Paperbound **$1.25**

EXPERIMENT AND THEORY IN PHYSICS, Max Born. A Nobel Laureate examines the nature of experiment and theory in theoretical physics and analyzes the advances made by the great physicists of our day: Heisenberg, Einstein, Bohr, Planck, Dirac, and others. The actual process of creation is detailed step-by-step by one who participated. A fine examination of the scientific method at work. 44pp. 5⅜ x 8.　　S308 Paperbound **75¢**

THE PSYCHOLOGY OF INVENTION IN THE MATHEMATICAL FIELD, J. Hadamard. The reports of such men as Descartes, Pascal, Einstein, Poincaré, and others are considered in this investigation of the method of idea-creation in mathematics and other sciences and the thinking process in general. How do ideas originate? What is the role of the unconscious? What is Poincaré's forgetting hypothesis? are some of the fascinating questions treated. A penetrating analysis of Einstein's thought processes concludes the book. xiii + 145pp. 5⅜ x 8.　　T107 Paperbound **$1.25**

THE NATURE OF LIGHT AND COLOUR IN THE OPEN AIR, M. Minnaert. Why are shadows sometimes blue, sometimes green, or other colors depending on the light and surroundings? What causes mirages? Why do multiple suns and moons appear in the sky? Professor Minnaert explains these unusual phenomena and hundreds of others in simple, easy-to-understand terms based on optical laws and the properties of light and color. No mathematics is required but artists, scientists, students, and everyone fascinated by these "tricks" of nature will find thousands of useful and amazing pieces of information. Hundreds of observational experiments are suggested which require no special equipment. 200 illustrations; 42 photos. xvi + 362pp. 5⅜ x 8.　　T196 Paperbound **$2.00**

*MATHEMATICS IN ACTION, O. G. Sutton. Everyone with a command of high school algebra will find this book one of the finest possible introductions to the application of mathematics to physical theory. Ballistics, numerical analysis, waves and wavelike phenomena, Fourier series, group concepts, fluid flow and aerodynamics, statistical measures, and meteorology are discussed with unusual clarity. Some calculus and differential equations theory is developed by the author for the reader's help in the more difficult sections. 88 figures. Index. viii + 236pp. 5⅜ x 8.　　T440 Clothbound **$3.50**

SOAP-BUBBLES: THEIR COLOURS AND THE FORCES THAT MOULD THEM, C. V. Boys. For continuing popularity and validity as scientific primer, few books can match this volume of easily-followed experiments, explanations. Lucid exposition of complexities of liquid films, surface tension and related phenomena, bubbles' reaction to heat, motion, music, magnetic fields. Experiments with capillary attraction, soap bubbles on frames, composite bubbles, liquid cylinders and jets, bubbles other than soap, etc. Wonderful introduction to scientific method, natural laws that have many ramifications in areas of modern physics. Only complete edition in print. New Introduction by S. Z. Lewin, New York University. 83 illustrations; 1 full-page color plate. xii + 190pp. 5⅜ x 8½.　　T542 Paperbound **95¢**

THE STORY OF X-RAYS FROM RÖNTGEN TO ISOTOPES, A. R. Bleich, M.D. This book, by a member of the American College of Radiology, gives the scientific explanation of x-rays, their applications in medicine, industry and art, and their danger (and that of atmospheric radiation) to the individual and the species. You learn how radiation therapy is applied against cancer, how x-rays diagnose heart disease and other ailments, how they are used to examine mummies for information on diseases of early societies, and industrial materials for hidden weaknesses. 54 illustrations show x-rays of flowers, bones, stomach, gears with flaws, etc. 1st publication. Index. xix + 186pp. 5⅜ x 8. T622 Paperbound **$1.35**

SPINNING TOPS AND GYROSCOPIC MOTION, John Perry. A classic elementary text of the dynamics of rotation — the behavior and use of rotating bodies such as gyroscopes and tops. In simple, everyday English you are shown how quasi-rigidity is induced in discs of paper, smoke rings, chains, etc., by rapid motions; why a gyrostat falls and why a top rises; precession; how the earth's motion affects climate; and many other phenomena. Appendix on practical use of gyroscopes. 62 figures. 128pp. 5⅜ x 8. T416 Paperbound **$1.00**

SNOW CRYSTALS, W. A. Bentley, M. J. Humphreys. For almost 50 years W. A. Bentley photographed snow flakes in his laboratory in Jericho, Vermont; in 1931 the American Meteorological Society gathered together the best of his work, some 2400 photographs of snow flakes, plus a few ice flowers, windowpane frosts, dew, frozen rain, and other ice formations. Pictures were selected for beauty and scientific value. A very valuable work to anyone in meteorology, cryology; most interesting to layman; extremely useful for artist who wants beautiful, crystalline designs. All copyright free. Unabridged reprint of 1931 edition. 2453 illustrations. 227pp. 8 x 10½. T287 Paperbound **$3.00**

A DOVER SCIENCE SAMPLER, edited by George Barkin. A collection of brief, non-technical passages from 44 Dover Books Explaining Science for the enjoyment of the science-minded browser. Includes work of Bertrand Russell, Poincaré, Laplace, Max Born, Galileo, Newton; material on physics, mathematics, metallurgy, anatomy, astronomy, chemistry, etc. You will be fascinated by Martin Gardner's analysis of the sincere pseudo-scientist, Moritz's account of Newton's absentmindedness, Bernard's examples of human vivisection, etc. Illustrations from the Diderot Pictorial Encyclopedia and De Re Metallica. 64 pages. **FREE**

THE STORY OF ATOMIC THEORY AND ATOMIC ENERGY, J. G. Feinberg. A broader approach to subject of nuclear energy and its cultural implications than any other similar source. Very readable, informal, completely non-technical text. Begins with first atomic theory, 600 B.C. and carries you through the work of Mendelejeff, Röntgen, Madame Curie, to Einstein's equation and the A-bomb. New chapter goes through thermonuclear fission, binding energy, other events up to 1959. Radioactive decay and radiation hazards, future benefits, work of Bohr, moderns, hundreds more topics. "Deserves special mention . . . not only authoritative but thoroughly popular in the best sense of the word," Saturday Review. Formerly, "The Atom Story." Expanded with new chapter. Three appendixes. Index. 34 illustrations. vii + 243pp. 5⅜ x 8. T625 Paperbound **$1.60**

THE STRANGE STORY OF THE QUANTUM, AN ACCOUNT FOR THE GENERAL READER OF THE GROWTH OF IDEAS UNDERLYING OUR PRESENT ATOMIC KNOWLEDGE, B. Hoffmann. Presents lucidly and expertly, with barest amount of mathematics, the problems and theories which led to modern quantum physics. Dr. Hoffmann begins with the closing years of the 19th century, when certain trifling discrepancies were noticed, and with illuminating analogies and examples takes you through the brilliant concepts of Planck, Einstein, Pauli, Broglie, Bohr, Schroedinger, Heisenberg, Dirac, Sommerfeld, Feynman, etc. This edition includes a new, long postscript carrying the story through 1958. "Of the books attempting an account of the history and contents of our modern atomic physics which have come to my attention, this is the best," H. Margenau, Yale University, in "American Journal of Physics." 32 tables and line illustrations. Index. 275pp. 5⅜ x 8. T518 Paperbound **$1.50**

SPACE AND TIME, E. Borel. Written by a versatile mathematician of world renown with his customary lucidity and precision, this introduction to relativity for the layman presents scores of examples, analogies, and illustrations that open up new ways of thinking about space and time. It covers abstract geometry and geographical maps, continuity and topology, the propagation of light, the special theory of relativity, the general theory of relativity, theoretical researches, and much more. Mathematical notes. 2 Indexes. 4 Appendices. 15 figures. xvi + 243pp. 5⅜ x 8. T592 Paperbound **$1.45**

FROM EUCLID TO EDDINGTON: A STUDY OF THE CONCEPTIONS OF THE EXTERNAL WORLD, Sir Edmund Whittaker. A foremost British scientist traces the development of theories of natural philosophy from the western rediscovery of Euclid to Eddington, Einstein, Dirac, etc. The inadequacy of classical physics is contrasted with present day attempts to understand the physical world through relativity, non-Euclidean geometry, space curvature, wave mechanics, etc. 5 major divisions of examination: Space; Time and Movement; the Concepts of Classical Physics; the Concepts of Quantum Mechanics; the Eddington Universe. 212pp. 5⅜ x 8. T491 Paperbound **$1.35**

*THE EVOLUTION OF SCIENTIFIC THOUGHT FROM NEWTON TO EINSTEIN, A. d'Abro. A detailed account of the evolution of classical physics into modern relativistic theory and the concomitant changes in scientific methodology. The breakdown of classical physics in the face of non-Euclidean geometry and the electromagnetic equations is carefully discussed and then an exhaustive analysis of Einstein's special and general theories of relativity and their implications is given. Newton, Riemann, Weyl, Lorentz, Planck, Maxwell, and many others are considered. A non-technical explanation of space, time, electromagnetic waves, etc. as understood today. "Model of semi-popular exposition," NEW REPUBLIC. 21 diagrams. 482pp. 5⅜ x 8.
T2 Paperbound **$2.25**

EINSTEIN'S THEORY OF RELATIVITY, Max Born. Nobel Laureate explains Einstein's special and general theories of relativity, beginning with a thorough review of classical physics in simple, non-technical language. Exposition of Einstein's work discusses concept of simultaneity, kinematics, relativity of arbitrary motions, the space-time continuum, geometry of curved surfaces, etc., steering middle course between vague popularizations and complex scientific presentations. 1962 edition revised by author takes into account latest findings, predictions of theory and implications for cosmology, indicates what is being sought in unified field theory. Mathematics very elementary, illustrative diagrams and experiments informative but simple. Revised 1962 edition. Revised by Max Born, assisted by Gunther Leibfried and Walter Biem. Index. 143 illustrations. vii + 376pp. 5⅜ x 8.
S769 Paperbound **$2.00**

PHILOSOPHY AND THE PHYSICISTS, L. Susan Stebbing. A philosopher examines the philosophical aspects of modern science, in terms of a lively critical attack on the ideas of Jeans and Eddington. Such basic questions are treated as the task of science, causality, determinism, probability, consciousness, the relation of the world of physics to the world of everyday experience. The author probes the concepts of man's smallness before an inscrutable universe, the tendency to idealize mathematical construction, unpredictability theorems and human freedom, the supposed opposition between 19th century determinism and modern science, and many others. Introduces many thought-stimulating ideas about the implications of modern physical concepts. xvi + 295pp. 5⅜ x 8.
T480 Paperbound **$1.65**

THE RESTLESS UNIVERSE, Max Born. A remarkably lucid account by a Nobel Laureate of recent theories of wave mechanics, behavior of gases, electrons and ions, waves and particles, electronic structure of the atom, nuclear physics, and similar topics. "Much more thorough and deeper than most attempts . . . easy and delightful," CHEMICAL AND ENGINEERING NEWS. Special feature: 7 animated sequences of 60 figures each showing such phenomena as gas molecules in motion, the scattering of alpha particles, etc. 11 full-page plates of photographs. Total of nearly 600 illustrations. 351pp. 6⅛ x 9¼.
T412 Paperbound **$2.00**

THE COMMON SENSE OF THE EXACT SCIENCES, W. K. Clifford. For 70 years a guide to the basic concepts of scientific and mathematical thought. Acclaimed by scientists and laymen alike, it offers a wonderful insight into concepts such as the extension of meaning of symbols, characteristics of surface boundaries, properties of plane figures, measurement of quantities, vectors, the nature of position, bending of space, motion, mass and force, and many others. Prefaces by Bertrand Russell and Karl Pearson. Critical introduction by James Newman. 130 figures. 249pp. 5⅜ x 8.
T61 Paperbound **$1.60**

MATTER AND LIGHT, THE NEW PHYSICS, Louis de Broglie. Non-technical explanations by a Nobel Laureate of electro-magnetic theory, relativity, matter, light and radiation, wave mechanics, quantum physics, philosophy of science, and similar topics. This is one of the simplest yet most accurate introductions to the work of men like Planck, Einstein, Bohr, and others. Only 2 of the 21 chapters require a knowledge of mathematics. 300pp. 5⅜ x 8.
T35 Paperbound **$1.85**

SCIENCE, THEORY AND MAN, Erwin Schrödinger. This is a complete and unabridged reissue of SCIENCE AND THE HUMAN TEMPERAMENT plus an additional essay: "What Is an Elementary Particle?" Nobel Laureate Schrödinger discusses such topics as nature of scientific method, the nature of science, chance and determinism, science and society, conceptual models for physical entities, elementary particles and wave mechanics. Presentation is popular and may be followed by most people with little or no scientific training. "Fine practical preparation for a time when laws of nature, human institutions . . . are undergoing a critical examination without parallel," Waldemar Kaempffert, N. Y. TIMES. 192pp. 5⅜ x 8.
T428 Paperbound **$1.35**

CONCERNING THE NATURE OF THINGS, Sir William Bragg. The Nobel Laureate physicist in his Royal Institute Christmas Lectures explains such diverse phenomena as the formation of crystals, how uranium is transmuted to lead, the way X-rays work, why a spinning ball travels in a curved path, the reason why bubbles bounce from each other, and many other scientific topics that are seldom explained in simple terms. No scientific background needed—book is easy enough that any intelligent adult or youngster can understand it. Unabridged. 32pp. of photos; 57 figures. xii + 232pp. 5⅜ x 8.
T31 Paperbound **$1.35**

*THE RISE OF THE NEW PHYSICS (formerly THE DECLINE OF MECHANISM), A. d'Abro. This authoritative and comprehensive 2 volume exposition is unique in scientific publishing. Written for intelligent readers not familiar with higher mathematics, it is the only thorough explanation in non-technical language of modern mathematical-physical theory. Combining both history and exposition, it ranges from classical Newtonian concepts up through the electronic theories of Dirac and Heisenberg, the statistical mechanics of Fermi, and Einstein's relativity theories. "A must for anyone doing serious study in the physical sciences," J. OF FRANKLIN INST. 97 illustrations. 991pp. 2 volumes.
T3 Vol. 1, Paperbound **$2.25**
T4 Vol. 2, Paperbound **$2.25**

SCIENCE AND HYPOTHESIS, Henri Poincaré. Creative psychology in science. How such concepts as number, magnitude, space, force, classical mechanics were developed and how the modern scientist uses them in his thought. Hypothesis in physics, theories of modern physics. Introduction by Sir James Larmor. "Few mathematicians have had the breadth of vision of Poincaré, and none is his superior in the gift of clear exposition," E. T. Bell. Index. 272pp. 5⅜ x 8.
S221 Paperbound **$1.35**

THE VALUE OF SCIENCE, Henri Poincaré. Many of the most mature ideas of the "last scientific universalist" conveyed with charm and vigor for both the beginning student and the advanced worker. Discusses the nature of scientific truth, whether order is innate in the universe or imposed upon it by man, logical thought versus intuition (relating to mathematics through the works of Weierstrass, Lie, Klein, Riemann), time and space (relativity, psychological time, simultaneity), Hertz's concept of force, interrelationship of mathematical physics to pure math, values within disciplines of Maxwell, Carnot, Mayer, Newton, Lorentz, etc. Index. iii + 147pp. 5⅜ x 8.
S469 Paperbound **$1.35**

THE SKY AND ITS MYSTERIES, E. A. Beet. One of the most lucid books on the mysteries of the universe; covers history of astronomy from earliest observations to modern theories of expanding universe, source of stellar energy, birth of planets, origin of moon craters, possibilities of life on other planets. Discusses effects of sunspots on weather; distance, age of stars; methods and tools of astronomers; much more. Expert and fascinating. "Eminently readable book," London Times. Bibliography. Over 50 diagrams, 12 full-page plates. Fold-out star map. Introduction. Index. 238pp. 5¼ x 7½.
T627 Clothbound **$3.50**

OUT OF THE SKY: AN INTRODUCTION TO METEORITICS, H. H. Nininger. A non-technical yet comprehensive introduction to the young science of meteoritics: all aspects of the arrival of cosmic matter on our planet from outer space and the reaction and alteration of this matter in the terrestrial environment. Essential facts and major theories presented by one of the world's leading experts. Covers ancient reports of meteors; modern systematic investigations; fireball clusters; meteorite showers; tektites; planetoidal encounters; etc. 52 full-page plates with over 175 photographs. 22 figures. Bibliography and references. Index. viii + 336pp. 5⅜ x 8.
T519 Paperbound **$1.85**

THE REALM OF THE NEBULAE, E. Hubble. One of great astronomers of our day records his formulation of concept of "island universes." Covers velocity-distance relationship; classification, nature, distances, general types of nebulae; cosmological theories. A fine introduction to modern theories for layman. No math needed. New introduction by A. Sandage. 55 illustrations, photos. Index. iv + 201pp. 5⅜ x 8.
S455 Paperbound **$1.50**

AN ELEMENTARY SURVEY OF CELESTIAL MECHANICS, Y. Ryabov. Elementary exposition of gravitational theory and celestial mechanics. Historical introduction and coverage of basic principles, including: the ecliptic, the orbital plane, the 2- and 3-body problems, the discovery of Neptune, planetary rotation, the length of the day, the shapes of galaxies, satellites (detailed treatment of Sputnik I), etc. First American reprinting of successful Russian popular exposition. Follow actual methods of astrophysicists with only high school math! Appendix. 58 figures. 165pp. 5⅜ x 8.
T756 Paperbound **$1.25**

GREAT IDEAS AND THEORIES OF MODERN COSMOLOGY, Jagjit Singh. Companion volume to author's popular "Great Ideas of Modern Mathematics" (Dover, $1.55). The best non-technical survey of post-Einstein attempts to answer perhaps unanswerable questions of origin, age of Universe, possibility of life on other worlds, etc. Fundamental theories of cosmology and cosmogony recounted, explained, evaluated in light of most recent data: Einstein's concepts of relativity, space-time; Milne's a priori world-system; astrophysical theories of Jeans, Eddington; Hoyle's "continuous creation;" contributions of dozens more scientists. A faithful, comprehensive critical summary of complex material presented in an extremely well-written text intended for laymen. Original publication. Index. xii + 276pp. 5⅜ x 8½.
T925 Paperbound **$1.85**

BASIC ELECTRICITY, Bureau of Naval Personnel. Very thorough, easily followed course in basic electricity for beginner, layman, or intermediate student. Begins with simplest definitions, presents coordinated, systematic coverage of basic theory and application: conductors, insulators, static electricity, magnetism, production of voltage, Ohm's law, direct current series and parallel circuits, wiring techniques, electromagnetism, alternating current, capacitance and inductance, measuring instruments, etc.; application to electrical machines such as alternating and direct current generators, motors, transformers, magnetic magnifiers, etc. Each chapter contains problems to test progress; answers at rear. No math needed beyond algebra. Appendices on signs, formulas, etc. 345 illustrations. 448pp. 7½ x 10.
S973 Paperbound **$3.00**

ELEMENTARY METALLURGY AND METALLOGRAPHY, A. M. Shrager. An introduction to common metals and alloys; stress is upon steel and iron, but other metals and alloys also covered. All aspects of production, processing, working of metals. Designed for student who wishes to enter metallurgy, for bright high school or college beginner, layman who wants background on extremely important industry. Questions, at ends of chapters, many microphotographs, glossary. Greatly revised 1961 edition. 195 illustrations, tables. ix + 389pp. 5⅜ x 8.
S138 Paperbound **$2.25**

BRIDGES AND THEIR BUILDERS, D. B. Steinman & S. R. Watson. Engineers, historians, and every person who has ever been fascinated by great spans will find this book an endless source of information and interest. Greek and Roman structures, Medieval bridges, modern classics such as the Brooklyn Bridge, and the latest developments in the science are retold by one of the world's leading authorities on bridge design and construction. BRIDGES AND THEIR BUILDERS is the only comprehensive and accurate semi-popular history of these important measures of progress in print. New, greatly revised, enlarged edition. 23 photos; 26 line-drawings. Index. xvii + 401pp. 5⅜ x 8. **T431 Paperbound $2.00**

FAMOUS BRIDGES OF THE WORLD, D. B. Steinman. An up-to-the-minute new edition of a book that explains the fascinating drama of how the world's great bridges came to be built. The author, designer of the famed Mackinac bridge, discusses bridges from all periods and all parts of the world, explaining their various types of construction, and describing the problems their builders faced. Although primarily for youngsters, this cannot fail to interest readers of all ages. 48 illustrations in the text. 23 photographs. 99pp. 6⅛ x 9¼. **T161 Paperbound $1.00**

HOW DO YOU USE A SLIDE RULE? by A. A. Merrill. A step-by-step explanation of the slide rule that presents the fundamental rules clearly enough for the non-mathematician to understand. Unlike most instruction manuals, this work concentrates on the two most important operations: multiplication and division. 10 easy lessons, each with a clear drawing, for the reader who has difficulty following other expositions. 1st publication. Index. 2 Appendices. 10 illustrations. 78 problems, all with answers. vi + 36 pp. 6⅛ x 9¼. **T62 Paperbound 60¢**

HOW TO CALCULATE QUICKLY, H. Sticker. A tried and true method for increasing your "number sense" — the ability to see relationships between numbers and groups of numbers. Addition, subtraction, multiplication, division, fractions, and other topics are treated through techniques not generally taught in schools: left to right multiplication, division by inspection, etc. This is not a collection of tricks that work only on special numbers, but a detailed well-planned course, consisting of over 9,000 problems that you can work in spare moments. It is excellent for anyone who is inconvenienced by slow computational skills. 5 or 10 minutes of this book daily will double or triple your calculation speed. 9,000 problems, answers. 256pp. 5⅜ x 8. **T295 Paperbound $1.00**

MATHEMATICAL FUN, GAMES AND PUZZLES, Jack Frohlichstein. A valuable service for parents of children who have trouble with math, for teachers in need of a supplement to regular upper elementary and junior high math texts (each section is graded—easy, average, difficult —for ready adaptation to different levels of ability), and for just anyone who would like to develop basic skills in an informal and entertaining manner. The author combines ten years of experience as a junior high school math teacher with a method that uses puzzles and games to introduce the basic ideas and operations of arithmetic. Stress on everyday uses of math: banking, stock market, personal budgets, insurance, taxes. Intellectually stimulating and practical, too. 418 problems and diversions with answers. Bibliography. 120 illustrations. xix + 306pp. 5⅝ x 8½. **T789 Paperbound $1.75**

GREAT IDEAS OF MODERN MATHEMATICS: THEIR NATURE AND USE, Jagjit Singh. Reader with only high school math will understand main mathematical ideas of modern physics, astronomy, genetics, psychology, evolution, etc. better than many who use them as tools, but comprehend little of their basic structure. Author uses his wide knowledge of non-mathematical fields in brilliant exposition of differential equations, matrices, group theory, logic, statistics, problems of mathematical foundations, imaginary numbers, vectors, etc. Original publication. 2 appendixes. 2 indexes. 65 illustr. 322pp. 5⅜ x 8. **S587 Paperbound $1.75**

THE UNIVERSE OF LIGHT, W. Bragg. Sir William Bragg, Nobel Laureate and great modern physicist, is also well known for his powers of clear exposition. Here he analyzes all aspects of light for the layman: lenses, reflection, refraction, the optics of vision, x-rays, the photoelectric effect, etc. He tells you what causes the color of spectra, rainbows, and soap bubbles, how magic mirrors work, and much more. Dozens of simple experiments are described. Preface. Index. 199 line drawings and photographs, including 2 full-page color plates. x + 283pp. 5⅜ x 8. **T538 Paperbound $1.85**

*INTRODUCTION TO SYMBOLIC LOGIC AND ITS APPLICATIONS, Rudolph Carnap. One of the clearest, most comprehensive, and rigorous introductions to modern symbolic logic, by perhaps its greatest living master. Not merely elementary theory, but demonstrated applications in mathematics, physics, and biology. Symbolic languages of various degrees of complexity are analyzed, and one constructed. "A creation of the rank of a masterpiece," Zentralblatt für Mathematik und Ihre Grenzgebiete. Over 300 exercises. 5 figures. Bibliography. Index. xvi + 241pp. 5⅜ x 8. **S453 Paperbound $1.85**

*HIGHER MATHEMATICS FOR STUDENTS OF CHEMISTRY AND PHYSICS, J. W. Mellor. Not abstract, but practical, drawing its problems from familiar laboratory material, this book covers theory and application of differential calculus, analytic geometry, functions with singularities, integral calculus, infinite series, solution of numerical equations, differential equations, Fourier's theorem and extensions, probability and the theory of errors, calculus of variations, determinants, etc. "If the reader is not familiar with this book, it will repay him to examine it," CHEM. & ENGINEERING NEWS. 800 problems. 189 figures. 2 appendices; 30 tables of integrals, probability functions, etc. Bibliography. xxi + 641pp. 5⅜ x 8. **S193 Paperbound $2.50**

THE FOURTH DIMENSION SIMPLY EXPLAINED, edited by Henry P. Manning. Originally written as entries in contest sponsored by "Scientific American," then published in book form, these 22 essays present easily understood explanations of how the fourth dimension may be studied, the relationship of non-Euclidean geometry to the fourth dimension, analogies to three-dimensional space, some fourth-dimensional absurdities and curiosities, possible measurements and forms in the fourth dimension. In general, a thorough coverage of many of the simpler properties of fourth-dimensional space. Multi-points of view on many of the most important aspects are valuable aid to comprehension. Introduction by Dr. Henry P. Manning gives proper emphasis to points in essays, more advanced account of fourth-dimensional geometry. 82 figures. 251pp. 5⅜ x 8.　　　　　　　　　T711 Paperbound **$1.35**

TRIGONOMETRY REFRESHER FOR TECHNICAL MEN, A. A. Klaf. A modern question and answer text on plane and spherical trigonometry. Part I covers plane trigonometry: angles, quadrants, trigonometrical functions, graphical representation, interpolation, equations, logarithms, solution of triangles, slide rules, etc. Part II discusses applications to navigation, surveying, elasticity, architecture, and engineering. Small angles, periodic functions, vectors, polar coordinates, De Moivre's theorem, fully covered. Part III is devoted to spherical trigonometry and the solution of spherical triangles, with applications to terrestrial and astronomical problems. Special time-savers for numerical calculation. 913 questions answered for you! 1738 problems; answers to odd numbers. 494 figures. 14 pages of functions, formulae. Index. x + 629pp. 5⅜ x 8.　　　　　　　　　　　　　　　　T371 Paperbound **$2.00**

CALCULUS REFRESHER FOR TECHNICAL MEN. A. A. Klaf. Not an ordinary textbook but a unique refresher for engineers, technicians, and students. An examination of the most important aspects of differential and integral calculus by means of 756 key questions. Part I covers simple differential calculus: constants, variables, functions, increments, derivatives, logarithms, curvature, etc. Part II treats fundamental concepts of integration: inspection, substitution, transformation, reduction, areas and volumes, mean value, successive and partial integration, double and triple integration. Stresses practical aspects! A 50 page section gives applications to civil and nautical engineering, electricity, stress and strain, elasticity, industrial engineering, and similar fields. 756 questions answered. 556 problems; solutions to odd numbers. 36 pages of constants, formulae. Index. v + 431pp. 5⅜ x 8.
　　　　　　　　　　　　　　　　　　　　　　　T370 Paperbound **$2.00**

PROBABILITIES AND LIFE, Emile Borel. One of the leading French mathematicians of the last 100 years makes use of certain results of mathematics of probabilities and explains a number of problems that for the most part, are related to everyday living or to illness and death: computation of life expectancy tables, chances of recovery from various diseases, probabilities of job accidents, weather predictions, games of chance, and so on. Emphasis on results not processes, though some indication is made of mathematical proofs. Simple in style, free of technical terminology, limited in scope to everyday situations, it is comprehensible to laymen, fine reading for beginning students of probability. New English translation. Index. Appendix. vi + 87pp. 5⅜ x 8½.　　　　　T121 Paperbound **$1.00**

POPULAR SCIENTIFIC LECTURES, Hermann von Helmholtz. 7 lucid expositions by a pre-eminent scientific mind: "The Physiological Causes of Harmony in Music," "On the Relation of Optics to Painting," "On the Conservation of Force," "On the Interaction of Natural Forces," "On Goethe's Scientific Researches" into theory of color, "On the Origin and Significance of Geometric Axioms," "On Recent Progress in the Theory of Vision." Written with simplicity of expression, stripped of technicalities, these are easy to understand and delightful reading for anyone interested in science or looking for an introduction to serious study of acoustics or optics. Introduction by Professor Morris Kline, Director, Division of Electromagnetic Research, New York University, contains astute, impartial evaluations. Selected from "Popular Lectures on Scientific Subjects," 1st and 2nd series. xii + 286pp. 5⅜ x 8½.　　　　　　　　　　　　　　　　　　　T799 Paperbound **$1.45**

SCIENCE AND METHOD, Henri Poincaré. Procedure of scientific discovery, methodology, experiment, idea-germination—the intellectual processes by which discoveries come into being. Most significant and most interesting aspects of development, application of ideas. Chapters cover selection of facts, chance, mathematical reasoning, mathematics, and logic; Whitehead, Russell, Cantor; the new mechanics, etc. 288pp. 5⅜ x 8.　　　S222 Paperbound **$1.50**

HEAT AND ITS WORKINGS, Morton Mott-Smith, Ph.D. An unusual book; to our knowledge the only middle-level survey of this important area of science. Explains clearly such important concepts as physiological sensation of heat and Weber's law, measurement of heat, evolution of thermometer, nature of heat, expansion and contraction of solids, Boyle's law, specific heat. BTU's and calories, evaporation, Andrews's isothermals, radiation, the relation of heat to light, many more topics inseparable from other aspects of physics. A wide, non-mathematical yet thorough explanation of basic ideas, theories, phenomena for laymen and beginning scientists illustrated by experiences of daily life. Bibliography. 50 illustrations. x + 165pp. 5⅜ x 8½.　　　　　　　　　　　　　　　T978 Paperbound **$1.00**

PHYSICS

General physics

FOUNDATIONS OF PHYSICS, R. B. Lindsay & H. Margenau. Excellent bridge between semi-popular works & technical treatises. A discussion ot methods of physical description, construction of theory; valuable to physicist with elementary calculus who is interested in ideas that give meaning to data, tools of modern physics. Contents include symbolism, mathematical equations; space & time foundations of mechanics; probability; physics & continua; electron theory; special & general relativity; quantum mechanics; causality. "Thorough and yet not overdetailed. Unreservedly recommended," NATURE (London). Unabridged, corrected edition. List of recommended readings. 35 illustrations. xi + 537pp. 5⅜ x 8.
S377 Paperbound **$2.75**

FUNDAMENTAL FORMULAS OF PHYSICS, ed. by D. H. Menzel. Highly useful, fully inexpensive reference and study text, ranging trom simple to highly sophisticated operations. Mathematics integrated into text—each cnapter stands as short textbook ot field represented. Voi. 1: Statistics, Physical Constants, Special Theory of Relativity, Hydrodynamics, Aerodynamics, Boundary Value Problems in Math. Physics; Viscosity, Electromagnetic Theory, etc. Vol. 2: Sound, Acoustics, Geometrical Optics, Electron Optics, High-Energy Phenomena, Magnetism, Biophysics, much more. Index. Total of 800pp. 5⅜ x 8.
Vol. 1 S595 Paperbound **$2.00**
Vol. 2 S596 Paperbound **$2.00**

MATHEMATICAL PHYSICS, D. H. Menzel. Thorough one-volume treatment of the mathematical techniques vital for classic mechanics, electromagnetic theory, quantum theory, and relativity. Written by the Harvard Professor of Astrophysics for junior, senior, and graduate courses, it gives clear explanations of all those aspects of function theory, vectors, matrices, dyadics, tensors, partial differential equations, etc., necessary tor the understanding of the various physical theories. Electron theory, relativity, and other topics seldom presented appear here in considerable detail. Scores of definitions, conversion factors, dimensional constants, etc. "More detailed than normal for an advanced text . . . excellent set of sections on Dyadics, Matrices, and Tensors," JOURNAL OF THE FRANKLIN INSTITUTE. Index. 193 problems, with answers. x + 412pp. 5⅜ x 8.
S56 Paperbound **$2.00**

THE SCIENTIFIC PAPERS OF J. WILLARD GIBBS. All the published papers of America's outstanding theoretical scientist (except for "Statistical Mechanics" and "Vector Analysis"). Vol I (thermodynamics) contains one of the most brilliant of all 19th-century scientific papers—the 300-page "On the Equilibrium of Heterogeneous Substances," which founded the science of physical chemistry, and clearly stated a number of highly important natural laws for the first time; 8 other papers complete the first volume. Vol II includes 2 papers on dynamics, 8 on vector analysis and multiple algebra, 5 on the electromagnetic theory of light, and 6 miscellaneous papers. Biographical sketch by H. A. Bumstead. Total of xxxvi + 718pp. 5⅝ x 8⅜.
S721 Vol I Paperbound **$2.50**
S722 Vol II Paperbound **$2.00**
The set **$4.50**

BASIC THEORIES OF PHYSICS, Peter Gabriel Bergmann. Two-volume set which presents a critical examination of important topics in the major subdivisions of classical and modern physics. The first volume is concerned with classical mechanics and electrodynamics: mechanics of mass points, analytical mechanics, matter in bulk, electrostatics and magnetostatics, electromagnetic interaction, the field waves, special relativity, and waves. The second volume (Heat and Quanta) contains discussions of the kinetic hypothesis, physics and statistics, stationary ensembles, laws of thermodynamics, early quantum theories, atomic spectra, probability waves, quantization in wave mechanics, approximation methods, and abstract quantum theory. A valuable supplement to any thorough course or text.
Heat and Quanta: Index. 8 figures. x + 300pp. 5⅜ x 8½.
S968 Paperbound **$2.00**
Mechanics and Electrodynamics: Index. 14 figures. vii + 280pp. 5⅜ x 8½.
S969 Paperbound **$1.75**

THEORETICAL PHYSICS, A. S. Kompaneyets. One of the very few thorough studies of the subject in this price range. Provides advanced students with a comprehensive theoretical background. Especially strong on recent experimentation and developments in quantum theory. Contents: Mechanics (Generalized Coordinates, Lagrange's Equation, Collision of Particles, etc.), Electrodynamics (Vector Analysis, Maxwell's equations, Transmission of Signals, Theory of Relativity, etc.), Quantum Mechanics (the Inadequacy of Classical Mechanics, the Wave Equation, Motion in a Central Field, Quantum Theory of Radiation, Quantum Theories of Dispersion and Scattering, etc.), and Statistical Physics (Equilibrium Distribution of Molecules in an Ideal Gas, Boltzmann statistics, Bose and Fermi Distribution, Thermodynamic Quantities, etc.). Revised to 1961. Translated by George Yankovsky, authorized by Kompaneyets. 137 exercises. 56 figures. 529pp. 5⅜ x 8½. S972 Paperbound **$2.50**

ANALYTICAL AND CANONICAL FORMALISM IN PHYSICS, André Mercier. A survey, in one volume, of the variational principles (the key principles—in mathematical form—from which the basic laws of any one branch of physics can be derived) of the several branches of physical theory, together with an examination of the relationships among them. Contents: the Lagrangian Formalism, Lagrangian Densities, Canonical Formalism, Canonical Form of Electrodynamics, Hamiltonian Densities, Transformations, and Canonical Form with Vanishing Jacobian Determinant. Numerous examples and exercises. For advanced students, teachers, etc. 6 figures. Index. viii + 222pp. 5⅜ x 8½.
S1077 Paperbound **$1.75**

CHEMISTRY AND PHYSICAL CHEMISTRY

ORGANIC CHEMISTRY, F. C. Whitmore. The entire subject of organic chemistry for the practic-
ing chemist and the advanced student. Storehouse of facts, theories, processes found elsewhere only in specialized journals. Covers aliphatic compounds (500 pages on the properties
and synthetic preparation of hydrocarbons, halides, proteins, ketones, etc.), alicyclic compounds, aromatic compounds, heterocyclic compounds, organophosphorus and organometallic
compounds. Methods of synthetic preparation analyzed critically throughout. Includes much of
biochemical interest. "The scope of this volume is astonishing," INDUSTRIAL AND ENGINEER-
ING CHEMISTRY. 12,000-reference index. 2387-item bibliography. Total of x + 1005pp. 5⅜ x 8.
Two volume set. S700 Vol I Paperbound **$2.25**
S701 Vol II Paperbound **$2.25**
The set **$4.50**

THE MODERN THEORY OF MOLECULAR STRUCTURE, Bernard Pullman. A reasonably popular
account of recent developments in atomic and molecular theory. Contents: The Wave Function and Wave Equations (history and bases of present theories of molecular structure);
The Electronic Structure of Atoms (Description and classification of atomic wave functions,
etc.); Diatomic Molecules; Non-Conjugated Polyatomic Molecules; Conjugated Polyatomic
Molecules; The Structure of Complexes. Minimum of mathematical background needed. New
translation by David Antin of "La Structure Moleculaire." Index. Bibliography. vii + 87pp.
5⅜ x 8½. S987 Paperbound **$1.00**

CATALYSIS AND CATALYSTS, Marcel Prettre, Director, Research Institute on Catalysis. This
brief book, translated into English for the first time, is the finest summary of the principal
modern concepts, methods, and results of catalysis. Ideal introduction for beginning chemistry and physics students. Chapters: Basic Definitions of Catalysis (true catalysis and
generalization of the concept of catalysis); The Scientific Bases of Catalysis (Catalysis
and chemical thermodynamics, catalysis and chemical kinetics); Homogeneous Catalysis
(acid-base catalysis, etc.); Chain Reactions; Contact Masses; Heterogeneous Catalysis
(Mechanisms of contact catalyses, etc.); and Industrial Applications (acids and fertilizers,
petroleum and petroleum chemistry, rubber, plastics, synthetic resins, and fibers). Translated by David Antin. Index. vi + 88pp. 5⅜ x 8½. S998 Paperbound **$1.00**

POLAR MOLECULES, Pieter Debye. This work by Nobel laureate Debye offers a complete guide
to fundamental electrostatic field relations, polarizability, molecular structure. Partial contents: electric intensity, displacement and force, polarization by orientation, molar polarization and molar refraction, halogen-hydrides, polar liquids, ionic saturation, dielectric constant, etc. Special chapter considers quantum theory. Indexed. 172pp. 5⅜ x 8.
S64 Paperbound **$1.65**

THE ELECTRONIC THEORY OF ACIDS AND BASES, W. F. Luder and Saverio Zuffanti. The first
full systematic presentation of the electronic theory of acids and bases—treating the
theory and its ramifications in an uncomplicated manner. Chapters: Historical Background;
Atomic Orbitals and Valence; The Electronic Theory of Acids and Bases; Electrophilic and
Electrodotic Reagents; Acidic and Basic Radicals; Neutralization; Titrations with Indicators;
Displacement; Catalysis; Acid Catalysis; Base Catalysis; Alkoxides and Catalysts; Conclusion. Required reading for all chemists. Second revised (1961) eidtion, with additional
examples and references. 3 figures. 9 tables. Index. Bibliography xii + 165pp. 5⅜ x 8.
S201 Paperbound **$1.50**

KINETIC THEORY OF LIQUIDS, J. Frenkel. Regarding the kinetic theory of liquids as a generalization and extension of the theory of solid bodies, this volume covers all types of
arrangements of solids, thermal displacements of atoms, interstitial atoms and ions,
orientational and rotational motion of molecules, and transition between states of matter.
Mathematical theory is developed close to the physical subject matter. 216 bibliographical
footnotes. 55 figures. xi + 485pp. 5⅜ x 8. S95 Paperbound **$2.55**

THE PRINCIPLES OF ELECTROCHEMISTRY, D. A. MacInnes. Basic equations for almost every
subfield of electrochemistry from first principles, referring at all times to the soundest and
most recent theories and results; unusually useful as text or as reference. Covers coulometers
and Faraday's Law, electrolytic conductance, the Debye-Hueckel method for the theoretical
calculation of activity coefficients, concentration cells, standard electrode potentials, thermodynamic ionization constants, pH, potentiometric titrations, irreversible phenomena, Planck's
equation, and much more. "Excellent treatise," AMERICAN CHEMICAL SOCIETY JOURNAL.
"Highly recommended," CHEMICAL AND METALLURGICAL ENGINEERING. 2 Indices. Appendix.
585-item bibliography. 137 figures. 94 tables. ii + 478pp. 5⅝ x 8⅜.
S52 Paperbound **$2.45**

THE PHASE RULE AND ITS APPLICATION, Alexander Findlay. Covering chemical phenomena
of 1, 2, 3, 4, and multiple component systems, this "standard work on the subject"
(NATURE, London), has been completely revised and brought up to date by A. N. Campbell
and N. O. Smith. Brand new material has been added on such matters as binary, tertiary
liquid equilibria, solid solutions in ternary systems, quinary systems of salts and water.
Completely revised to triangular coordinates in ternary systems, clarified graphic representation, solid models, etc. 9th revised edition. Author, subject indexes. 236 figures. 505
footnotes, mostly bibliographic. xii + 494pp. 5⅜ x 8. S91 Paperbound **$2.50**

Social Sciences

SOCIAL THOUGHT FROM LORE TO SCIENCE, H. E. Barnes and H. Becker. An immense survey of sociological thought and ways of viewing, studying, planning, and reforming society from earliest times to the present. Includes thought on society of preliterate peoples, ancient non-Western cultures, and every great movement in Europe, America, and modern Japan. Analyzes hundreds of great thinkers: Plato, Augustine, Bodin, Vico, Montesquieu, Herder, Comte, Marx, etc. Weighs the contributions of utopians, sophists, fascists and communists; economists, jurists, philosophers, ecclesiastics, and every 19th and 20th century school of scientific sociology, anthropology, and social psychology throughout the world. Combines topical, chronological, and regional approaches, treating the evolution of social thought as a process rather than as a series of mere topics. "Impressive accuracy, competence, and discrimination . . . easily the best single survey," Nation. Thoroughly revised, with new material up to 1960. 2 indexes. Over 2200 bibliographical notes. Three volume set. Total of 1586pp. 5⅜ x 8.

T901 Vol I   Paperbound **$2.50**
T902 Vol II Paperbound **$2.50**
T903 Vol III Paperbound **$2.50**
The set **$7.50**

FOLKWAYS, William Graham Sumner. A classic of sociology, a searching and thorough examination of patterns of behaviour from primitive, ancient Greek and Judaic, Medieval Christian, African, Oriental, Melanesian, Australian, Islamic, to modern Western societies. Thousands of illustrations of social, sexual, and religious customs, mores, laws, and institutions. Hundreds of categories: Labor, Wealth, Abortion, Primitive Justice, Life Policy, Slavery, Cannibalism, Uncleanness and the Evil Eye, etc. Will extend the horizon of every reader by showing the relativism of his own culture. Prefatory note by A. G. Keller. Introduction by William Lyon Phelps. Bibliography. Index. xiii + 692pp. 5⅜ x 8.   T508 Paperbound **$2.49**

PRIMITIVE RELIGION, P. Radin. A thorough treatment by a noted anthropologist of the nature and origin of man's belief in the supernatural and the influences that have shaped religious expression in primitive societies. Ranging from the Arunta, Ashanti, Aztec, Bushman, Crow, Fijian, etc., of Africa, Australia, Pacific Islands, the Arctic, North and South America, Prof. Radin integrates modern psychology, comparative religion, and economic thought with first-hand accounts gathered by himself and other scholars of primitive initiations, training of the shaman, and other fascinating topics. "Excellent," NATURE (London). Unabridged reissue of 1st edition. New author's preface. Bibliographic notes. Index. x + 322pp. 5⅜ x 8.
T393 Paperbound **$2.00**

PRIMITIVE MAN AS PHILOSOPHER, P. Radin. A standard anthropological work covering primitive thought on such topics as the purpose of life, marital relations, freedom of thought, symbolism, death, resignation, the nature of reality, personality, gods, and many others. Drawn from factual material gathered from the Winnebago, Oglala Sioux, Maori, Baganda, Batak, Zuni, among others, it does not distort ideas by removing them from context but interprets strictly within the original framework. Extensive selections of original primitive documents. Bibliography. Index. xviii + 402pp. 5⅜ x 8.   T392 Paperbound **$2.25**

A TREATISE ON SOCIOLOGY, THE MIND AND SOCIETY, Vilfredo Pareto. This treatise on human society is one of the great classics of modern sociology. First published in 1916, its careful catalogue of the innumerable manifestations of non-logical human conduct (Book One); the theory of "residues," leading to the premise that sentiment not logic determines human behavior (Book Two), and of "derivations," beliefs derived from desires (Book Three); and the general description of society made up of non-elite and elite, consisting of "foxes" who live by cunning and "lions" who live by force, stirred great controversy. But Pareto's passion for isolation and classification of elements and factors, and his allegiance to scientific method as the key tool for scrutinizing the human situation made his a truly twentieth-century mind and his work a catalytic influence on certain later social commentators. These four volumes (bound as two) require no special training to be appreciated and any reader who wishes to gain a complete understanding of modern sociological theory, regardless of special field of interest, will find them a must. Reprint of revised (corrected) printing of original edition. Translated by Andrew Bongiorno and Arthur Livingston. Index. Bibliography. Appendix containing index-summary of theorems. 48 diagrams. Four volumes bound as two. Total of 2063pp. 5⅜ x 8½.   The set Clothbound **$15.00**

THE POLISH PEASANT IN EUROPE AND AMERICA, William I. Thomas, Florian Znaniecki. A seminal sociological study of peasant primary groups (family and community) and the disruptions produced by a new industrial system and immigration to America. The peasant's family, class system, religious and aesthetic attitudes, and economic life are minutely examined and analyzed in hundreds of pages of primary documentation, particularly letters between family members. The disorientation caused by new environments is scrutinized in detail (a 312-page autobiography of an immigrant is especially valuable and revealing) in an attempt to find common experiences and reactions. The famous "Methodological Note" sets forth the principles which guided the authors. When out of print this set has sold for as much as $50. 2nd revised edition. 2 vols. Vol. 1: xv + 1115pp. Vol. 2: 1135pp. Index. 6 x 9.
T478 Clothbound 2 vol. set **$12.50**

Dover Classical Records

Now available directly to the public exclusively from Dover: top-quality recordings of fine classical music for only $2 per record! Originally released by a major company (except for the previously unreleased Gimpel recording of Bach) to sell for $5 and $6, these records were issued under our imprint only after they had passed a severe critical test. We insisted upon:

First-rate music that is enjoyable, musically important and culturally significant.

First-rate performances, where the artists have carried out the composer's intentions, in which the music is alive, vigorous, played with understanding and sympathy.

First-rate sound—clear, sonorous, fully balanced, crackle-free, whir-free.

Have in your home music by major composers, performed by such gifted musicians as Elsner, Gitlis, Wührer, the Barchet Quartet, Gimpel. Enthusiastically received when first released, many of these performances are definitive. The records are not seconds or remainders, but brand new pressings made on pure vinyl from carefully chosen master tapes. "All purpose" 12" monaural 33⅓ rpm records, they play equally well on hi-fi and stereo equipment. Fine music for discriminating music lovers, superlatively played, flawlessly recorded: there is no better way to build your library of recorded classical music at remarkable savings. There are no strings; this is not a come-on, not a club, forcing you to buy records you may not want in order to get a few at a lower price. Buy whatever records you want in any quantity, and never pay more than $2 each. Your obligation ends with your first purchase. And that's when ours begins. Dover's money-back guarantee allows you to return any record for any reason, even if you don't like the music, for a full, immediate refund, no questions asked.

MOZART: STRING QUARTET IN A MAJOR (K.464); STRING QUARTET IN C MAJOR ("DISSONANT", K.465), Barchet Quartet. The final two of the famed Haydn Quartets, high-points in the history of music. The A Major was accepted with delight by Mozart's contemporaries, but the C Major, with its dissonant opening, aroused strong protest. Today, of course, the remarkable resolutions of the dissonances are recognized as major musical achievements. "Beautiful warm playing," MUSICAL AMERICA. "Two of Mozart's loveliest quartets in a distinguished performance," REV. OF RECORDED MUSIC. (Playing time 58 mins.) HCR 5200 **$2.00**

MOZART: QUARTETS IN G MAJOR (K.80); D MAJOR (K.155); G MAJOR (K.156); C MAJOR (K157), Barchet Quartet. The early chamber music of Mozart receives unfortunately little attention. First-rate music of the Italian school, it contains all the lightness and charm that belongs only to the youthful Mozart. This is currently the only separate source for the composer's work of this time period. "Excellent," HIGH FIDELITY. "Filled with sunshine and youthful joy; played with verve, recorded sound live and brilliant," CHRISTIAN SCI. MONITOR. (Playing time 51 mins.) HCR 5201 **$2.00**

MOZART: SERENADE #9 IN D MAJOR ("POSTHORN", K.320); SERENADE #6 IN D MAJOR ("SERENATA NOTTURNA", K.239), Pro Musica Orch. of Stuttgart, under Edouard van Remoortel. For Mozart, the serenade was a highly effective form, since he could bring to it the immediacy and intimacy of chamber music as well as the free fantasy of larger group music. Both these serenades are distinguished by a playful, mischievous quality, a spirit perfectly captured in this fine performance. "A triumph, polished playing from the orchestra," HI FI MUSIC AT HOME. "Sound is rich and resonant, fidelity is wonderful," REV. OF RECORDED MUSIC. (Playing time 51 mins.) HCR 5202 **$2.00**

MOZART: DIVERTIMENTO IN E FLAT MAJOR FOR STRING TRIO (K.563); ADAGIO AND FUGUE IN F MINOR FOR STRING TRIO (K.404a), Kehr Trio. The Divertimento is one of Mozart's most beloved pieces, called by Einstein "the finest, most perfect trio ever heard." It is difficult to imagine a music lover who will not be delighted by it. This is the only recording of the lesser known Adagio and Fugue, written in 1782 and influenced by Bach's Well-Tempered Clavichord. "Extremely beautiful recording, strongly recommended," THE OBSERVER. "Superior to rival editions," HIGH FIDELITY. (Playing time 51 mins.) HCR 5203 **$2.00**

SCHUMANN: KREISLERIANA (OP.16); FANTASY IN C MAJOR ("FANTASIE," OP.17), Vlado Perlemuter, Piano. The vigorous Romantic imagination and the remarkable emotional qualities of Schumann's piano music raise it to special eminence in 19th century creativity. Both these pieces are rooted to the composer's tortuous romance with his future wife, Clara, and both receive brilliant treatment at the hands of Vlado Perlemuter, Paris Conservatory, proclaimed by Alfred Cortot "not only a great virtuoso but also a great musician." "The best Kreisleriana to date," BILLBOARD. (Playing time 55 mins.) HCR 5204 **$2.00**

SCHUMANN: TRIO #1, D MINOR; TRIO #3, G MINOR, Trio di Bolzano. The fiery, romantic, melodic Trio #1, and the dramatic, seldom heard Trio #3 are both movingly played by a fine chamber ensemble. No one personified Romanticism to the general public of the 1840's more than did Robert Schumann, and among his most romantic works are these trios for cello, violin and piano. "Ensemble and overall interpretation leave little to be desired," HIGH FIDELITY. "An especially understanding performance," REV. OF RECORDED MUSIC. (Playing time 54 mins.) HCR 5205 **$2.00**

SCHUMANN: TRIOS #1 IN D MINOR (OPUS 63) AND #3 IN G MINOR (OPUS 110), Trio di Bolzano. The fiery, romantic, melodic Trio #1 and the dramatic, seldom heard Trio #3 are both movingly played by a fine chamber ensemble. No one personified Romanticism to the general public of the 1840's more than did Robert Schumann, and among his most romantic works are these trios for cello, violin and piano. "Ensemble and overall interpretation leave little to be desired," HIGH FIDELITY. "An especially understanding performance," REV. OF RECORDED MUSIC. (Playing time 54 mins.)   HCR 5205 **$2.00**

SCHUBERT: QUINTET IN A ("TROUT") (OPUS 114), AND NOCTURNE IN E FLAT (OPUS 148), Friedrich Wührer, Piano and Barchet Quartet. If there is a single piece of chamber music that is a universal favorite, it is probably Schubert's "Trout" Quintet. Delightful melody, harmonic resources, musical exuberance are its characteristics. The Nocturne (played by Wührer, Barchet, and Reimann) is an exquisite piece with a deceptively simple theme and harmony. "The best Trout on the market—Wührer is a fine Viennese-style Schubertian, and his spirit infects the Barchets," ATLANTIC MONTHLY. "Exquisitely recorded," ETUDE. (Playing time 44 mins.)   HCR 5206 **$2.00**

SCHUBERT: PIANO SONATAS IN C MINOR AND B (OPUS 147), Friedrich Wührer. Schubert's sonatas retain the structure of the classical form, but delight listeners with romantic freedom and a special melodic richness. The C Minor, one of the Three Grand Sonatas, is a product of the composer's maturity. The B Major was not published until 15 years after his death. "Remarkable interpretation, reproduction of the first rank," DISQUES. "A superb pianist for music like this, musicianship, sweep, power, and an ability to integrate Schubert's measures such as few pianists have had since Schnabel," Harold Schonberg. (Playing time 49 mins.)   HCR 5207 **$2.00**

STRAVINSKY: VIOLIN CONCERTO IN D, Ivry Gitlis, Cologne Orchestra; DUO CONCERTANTE, Ivry Gitlis, Violin, Charlotte Zelka, Piano, Cologne Orchestra; JEU DE CARTES, Bamberg Symphony, under Hollreiser. Igor Stravinsky is probably the most important composer of this century, and these three works are among the most significant of his neoclassical period of the 30's. The Violin Concerto is one of the few modern classics. Jeu de Cartes, a ballet score, bubbles with gaiety, color and melodiousness. "Imaginatively played and beautifully recorded," E. T. Canby, HARPERS MAGAZINE. "Gitlis is excellent, Hollreiser beautifully worked out," HIGH FIDELITY. (Playing time 55 mins.)   HCR 5208 **$2.00**

GEMINIANI: SIX CONCERTI GROSSI, OPUS 3, Helma Elsner, Harpsichord, Barchet Quartet, Pro Musica Orch. of Stuttgart, under Reinhardt. Francesco Geminiani (1687-1762) has been rediscovered in the same musical exploration that revealed Scarlatti, Vivaldi, and Corelli. In form he is more sophisticated than the earlier Italians, but his music delights modern listeners with its combination of contrapuntal techniques and the full harmonies and rich melodies charcteristic of Italian music. This is the only recording of the six 1733 concerti: D Major, B Flat Minor, E Minor, G Minor, E Minor (bis), and D Minor. "I warmly recommend it, spacious, magnificent, I enjoyed every bar," C. Cudworth, RECORD NEWS. "Works of real charm, recorded with understanding and style," ETUDE. (Playing time 52 mins.)
                                                                       HCR 5209 **$2.00**

MODERN PIANO SONATAS: BARTOK: SONATA FOR PIANO; BLOCH: SONATA FOR PIANO (1935); PROKOFIEV, PIANO SONATA #7 IN B FLAT ("STALINGRAD"); STRAVINSKY: PIANO SONATA (1924), István Nádas, Piano. Shows some of the major forces and directions in modern piano music: Stravinsky's crisp austerity; Bartok's fusion of Hungarian folk motives; incisive diverse rhythms, and driving power; Bloch's distinctive emotional vigor; Prokofiev's brilliance and melodic beauty couched in pre-Romantic forms. "A most interesting documentation of the contemporary piano sonata. Nadas is a very good pianist." HIGH FIDELITY. (Playing time 59 mins.)   HCR 5215 **$2.00**

VIVALDI: CONCERTI FOR FLUTE, VIOLIN, BASSOON, AND HARPSICHORD: #8 IN G MINOR, #21 IN F, #27 IN D, #7 IN D; SONATA #1 IN A MINOR, Gastone Tassinari, Renato Giangrandi, Giorgio Semprini, Arlette Eggmann. More than any other Baroque composer, Vivaldi moved the concerto grosso closer to the solo concert we deem standard today. In these concerti he wrote virtuosi music for the solo instruments, allowing each to introduce new material or expand on musical ideas, creating tone colors unusual even for Vivaldi. As a result, this record displays a new area of his genius, offering some of his most brilliant music. Performed by a top-rank European group. (Playing time 45 mins.)   HCR 5216 **$2.00**

LÜBECK: CANTATAS: HILF DEINEM VOLK; GOTT, WIE DEIN NAME, Stuttgart Choral Society, Swabian Symphony Orch.; PRELUDES AND FUGUES IN C MINOR AND IN E, Eva Hölderlin, Organ. Vincent Lübeck (1654-1740), contemporary of Bach and Buxtehude, was one of the great figures of the 18th-century North German school. These examples of Lübeck's few surviving works indicate his power and brilliance. Voice and instrument lines in the cantatas are strongly reminiscent of the organ: the preludes and fugues show the influence of Bach and Buxtehude. This is the only recording of the superb cantatas. Text and translation included. "Outstanding record," E. T. Canby, SAT. REVIEW. "Hölderlin's playing is exceptional," AM. RECORD REVIEW. "Will make [Lübeck] many new friends," Philip Miller. (Playing time 37 mins.)   HCR 5217 **$2.00**

DONIZETTI: BETLY (LA CAPANNA SVIZZERA), Soloists of Compagnia del Teatro dell'Opera Comica di Roma, Societa del Quartetto, Rome, Chorus and Orch. Betly, a delightful one-act opera written in 1836, is similar in style and story to one of Donizetti's better-known operas, L'Elisir. Betly is lighthearted and farcical, with bright melodies and a freshness character-istic of the best of Donizetti. Libretto (English and Italian) included. "The chief honors go to Angela Tuccari who sings the title role, and the record is worth having for her alone," M. Rayment, GRAMOPHONE REC. REVIEW. "The interpretation . . . is excellent . . . This is a charming record which we recommend to lovers of little-known works," DISQUES.
HCR 5218 **$2.00**

ROSSINI: L'OCCASIONE FA IL LADRO (IL CAMBIO DELLA VALIGIA), Soloists of Compagnia del Teatro dell'Opera Comica di Roma, Societa del Quartetto, Rome, Chorus and Orch. A charm-ing one-act opera buffa, this is one of the first works of Rossini's maturity, and it is filled with the wit, gaiety and sparkle that make his comic operas second only to Mozart's. Like other Rossini works, L'Occasione makes use of the theme of impersonation and attendant amusing confusions. This is the only recording of this important buffa. Full libretto (English and Italian) included. "A major rebirth, a stylish performance . . . the Roman recording engineers have outdone themselves," H. Weinstock, SAT. REVIEW. (Playing time 53 mins.)
HCR 5219 **$2.00**

DOWLAND: "FIRST BOOKE OF AYRES," Pro Musica Antiqua of Brussels, Safford Cape, Director. This is the first recording to include all 22 of the songs of this great collection, written by John Dowland, one of the most important writers of songs of 16th and 17th century Eng-land. The participation of the Brussels Pro Musica under Safford Cape insures scholarly ac-curacy and musical artistry. "Powerfully expressive and very beautiful," B. Haggin. "The musicianly singers . . . never fall below an impressive standard," Philip Miller. Text included. (Playing time 51 mins.)
HCR 5220 **$2.00**

FRENCH CHANSONS AND DANCES OF THE 16TH CENTURY, Pro Musica Antiqua of Brussels, Safford Cape, Director. A remarkable selection of 26 three- or four-part chansons and de-lightful dances from the French Golden Age—by such composers as Orlando Lasso, Crecquil-lon, Claude Gervaise, etc. Text and translation included. "Delightful, well-varied with respect to mood and to vocal and instrumental color," HIGH FIDELITY. "Performed with . . . dis-crimination and musical taste, full of melodic distinction and harmonic resource," Irving Kolodin. (Playing time 39 mins.)
HCR 5221 **$2.00**

GALUPPI: CONCERTI A QUATRO: #1 IN G MINOR, #2 IN G, #3 IN D, #4 IN C MINOR, #5 IN E FLAT, AND #6 IN B FLAT, Biffoli Quartet. During Baldassare Galuppi's lifetime, his instru-mental music was widely renowned, and his contemporaries Mozart and Haydn thought highly of his work. These 6 concerti reflect his great ability; and they are among the most interesting compositions of the period. They are remarkable for their unusual combinations of timbres and for emotional elements that were only then beginning to be introduced into music. Performed by the well-known Biffoli Quartet, this is the only record devoted exclu-sively to Galuppi. (Playing time 47 mins.)
HCR 5222 **$2.00**

HAYDN: DIVERTIMENTI FOR WIND BAND, IN C; IN F; DIVERTIMENTO A NOVE STROMENTI IN C FOR STRINGS AND WIND INSTRUMENTS, reconstructed by H. C. Robbins Landon, performed by members of Vienna State Opera Orch.; MOZART DIVERTIMENTI IN C, III (K. 187) AND IV (K. 188), Salzburg Wind Ensemble. Robbins Landon discovered Haydn manuscripts in a Bene-dictine monastery in Lower Austria, edited them and restored their original instrumentation The result is this magnificent record. Two little-known divertimenti by Mozart—of great charm and appeal—are also included. None of this music is available elsewhere (Playing time 58 mins.)
HCR 5223 **$2.00**

PURCELL: TRIO SONATAS FROM "SONATAS OF FOUR PARTS" (1697): #9 IN F ("GOLDEN"), #7 IN C, #1 IN B MINOR, #10 IN D, #4 IN D MINOR, #2 IN E FLAT, AND #8 IN G MINOR, Giorgio Ciompi, and Werner Torkanowsky, Violins, Geo. Koutzen, Cello, and Herman Chessid, Harpsichord. These posthumously-published sonatas show Purcell at his most advanced and mature. They are certainly among the finest musical examples of pre-modern chamber music. Those not familiar with his instrumental music are well-advised to hear these outstanding pieces. "Performance sounds excellent," Harold Schonberg. "Some of the most noble and touching music known to anyone," AMERICAN RECORD GUIDE. (Playing time 58 mins.)
HCR 5224 **$2.00**

BARTOK: VIOLIN CONCERTO; SONATA FOR UNACCOMPANIED VIOLIN, Ivry Gitlis, Pro Musica of Vienna, under Hornstein. Both these works are outstanding examples of Bartok's final period, and they show his powers at their fullest. The Violin Concerto is, in the opinion of many authorities, Bartok's finest work, and the Sonata, his last work, is "a masterpiece" (F. Sackville West). "Wonderful, finest performance of both Bartok works I have ever heard," GRAMOPHONE. "Gitlis makes such potent and musical sense out of these works that I suspect many general music lovers (not otherwise in sympathy with modern music) will discover to their amazement that they like it. Exceptionally good sound," AUDITOR. (Playing time 54 mins.)
HCR 5211 **$2.00**

J. S. BACH: PARTITAS FOR UNACCOMPANIED VIOLIN: #2 in D Minor and #3 in E, Bronislav Gimpel. Bach's works for unaccompanied violin fall within the same area that produced the Brandenburg Concerti, the Orchestral Suites, and the first part of the Well-Tempered Clavichord. The D Minor is considered one of Bach's masterpieces; the E Major is a buoyant work with exceptionally interesting bariolage effects. This is the first release of a truly memorable recording by Bronislav Gimpel, "as a violinist, the equal of the greatest" (P. Leron, in OPERA, Paris). (Playing time 53 mins.)   HCR 5212 **$2.00**

ROSSINI: QUARTETS FOR WOODWINDS: #1 IN F, #4 IN B FLAT, #5 IN D, AND #6 IN F, N. Y. Woodwind Quartet Members: S. Baron, Flute, J. Barrows, French Horn; B. Garfield, Bassoon; D. Glazer, Clarinet. Rossini's great genius was centered in the opera, but he also wrote a small amount of first-rate non-vocal music. Among these instrumental works, first place is usually given to the very interesting quartets. Of the three different surviving arrangements, this wind group version is the original, and this is the first recording of these works. "Each member of the group displays wonderful virtuosity when the music calls for it, at other times blending sensitively into the ensemble," HIGH FIDELITY. "Sheer delight," Philip Miller. (Playing time 45 mins.)   HCR 5214 **$2.00**

TELEMANN: THE GERMAN FANTASIAS FOR HARPSICHORD (#1-12), Helma Elsner. Until recently, Georg Philip Telemann (1681-1767) was one of the mysteriously neglected great men of music. Recently he has received the attention he deserved. He created music that delights modern listeners with its freshness and originality. These fantasias are free in form and reveal the intricacy of thorough bass music, the harmonic wealth of the "new music," and a distinctive melodic beauty. "This is another blessing of the contemporary LP output. Miss Elsner plays with considerable sensitivity and a great deal of understanding," REV. OF RE-CORDED MUSIC. "Fine recorded sound," Harold Schonberg. "Recommended warmly, very high quality," DISQUES. (Playing time 50 mins.)   HCR 5210 **$2.00**

Nova Recordings

In addition to our reprints of outstanding out-of-print records and American releases of first-rate foreign recordings, we have established our own new records. In order to keep every phase of their production under our own control, we have engaged musicians of world renown to play important music (for the most part unavailable elsewhere), have made use of the finest recording studios in New York, and have produced tapes equal to anything on the market, we believe. The first of these entirely new records are now available.

RAVEL: GASPARD DE LA NUIT, LE TOMBEAU DE COUPERIN, JEUX D'EAU, Beveridge Webster, Piano. Webster studied under Ravel and played his works in European recitals, often with Ravel's personal participation in the program. This record offers examples of the three major periods of Ravel's pianistic work, and is a must for any serious collector or music lover. (Playing time about 50 minutes).   Monaural HCR 5213 **$2.00**
                                Stereo HCR ST 7000 **$2.00**

EIGHTEENTH CENTURY FRENCH FLUTE MUSIC, Jean-Pierre Rampal, Flute, and Robert Veyron-Lacroix, Harpsichord. Contains Concerts Royaux #7 for Flute and Harpsichord in G Minor, Francois Couperin; Sonata dite l'Inconnue in G for Flute and Harpsichord, Michel de la Barre; Sonata #6 in A Minor, Michel Blavet; and Sonata in D Minor, Anne Danican-Philidor. In the opinion of many Rampal is the world's premier flutist. (Playing time about 45 minutes)
                                Monaural   HCR 5238 **$2.00**
                                Stereo HCR ST 7001 **$2.00**

SCHUMANN: NOVELLETTEN (Opus 21), Beveridge Webster, Piano. Brilliantly played in this original recording by one of America's foremost keyboard performers. Connected Romantic pieces. Long a piano favorite. (Playing time about 45 minutes)
                                Monaural HCR 5239 **$2.00**
                                Stereo HCR ST 7002 **$2.00**

Language Books and Records

GERMAN: HOW TO SPEAK AND WRITE IT. AN INFORMAL CONVERSATIONAL METHOD FOR SELF STUDY, Joseph Rosenberg. Eminently useful for self study because of concentration on elementary stages of learning. Also provides teachers with remarkable variety of aids: 28 full- and double-page sketches with pertinent items numbered and identified in German and English; German proverbs, jokes; grammar, idiom studies; extensive practice exercises. The most interesting introduction to German available, full of amusing illustrations, photographs of cities and landmarks in German-speaking cities, cultural information subtly woven into conversational material. Includes summary of grammar, guide to letter writing, study guide to German literature by Dr. Richard Friedenthal. Index. 400 illustrations. 384pp. 5⅜ x 8½.
T271 Paperbound **$2.00**

FRENCH: HOW TO SPEAK AND WRITE IT. AN INFORMAL CONVERSATIONAL METHOD FOR SELF STUDY, Joseph Lemaitre. Even the absolute beginner can acquire a solid foundation for further study from this delightful elementary course. Photographs, sketches and drawings, sparkling colloquial conversations on a wide variety of topics (including French culture and custom), French sayings and quips, are some of aids used to demonstrate rather than merely describe the language. Thorough yet surprisingly entertaining approach, excellent for teaching and for self study. Comprehensive analysis of pronunciation, practice exercises and appendices of verb tables, additional vocabulary, other useful material. Index. Appendix. 400 illustrations. 416pp. 5⅜ x 8½.
T268 Paperbound **$2.00**

DICTIONARY OF SPOKEN SPANISH, Spanish-English, English-Spanish. Compiled from spoken Spanish, emphasizing idiom and colloquial usage in both Castilian and Latin-American. More than 16,000 entries containing over 25,000 idioms—the largest list of idiomatic constructions ever published. Complete sentences given, indexed under single words—language in immediately useable form, for travellers, businessmen, students, etc. 25 page introduction provides rapid survey of sounds, grammar, syntax, with full consideration of irregular verbs. Especially apt in modern treatment of phrases and structure. 17 page glossary gives translations of geographical names, money values, numbers, national holidays, important street signs, useful expressions of high frequency, plus unique 7 page glossary of Spanish and Spanish-American foods and dishes. Originally published as War Department Technical Manual TM 30-900. iv + 513pp. 5⅜ x 8.
T495 Paperbound **$1.75**

SPEAK MY LANGUAGE: SPANISH FOR YOUNG BEGINNERS, M. Ahlman, Z. Gilbert. Records provide one of the best, and most entertaining, methods of introducing a foreign language to children. Within the framework of a train trip from Portugal to Spain, an English-speaking child is introduced to Spanish by a native companion. (Adapted from a successful radio program of the N. Y. State Educational Department.) Though a continuous story, there are a dozen specific categories of expressions, including greetings, numbers, time, weather, food, clothes, family members, etc. Drill is combined with poetry and contextual use. Authentic background music is heard. An accompanying book enables a reader to follow the records, and includes a vocabulary of over 350 recorded expressions. Two 10″ 33⅓ records, total of 40 minutes. Book. 40 illustrations. 69pp. 5¼ x 10½.
T890 The set **$4.95**

AN ENGLISH-FRENCH-GERMAN-SPANISH WORD FREQUENCY DICTIONARY, H. S. Eaton. An indispensable language study aid, this is a semantic frequency list of the 6000 most frequently used words in 4 languages—24,000 words in all. The lists, based on concepts rather than words alone, and containing all modern, exact, and idiomatic vocabulary, are arranged side by side to form a unique 4-language dictionary. A simple key indicates the importance of the individual words within each language. Over 200 pages of separate indexes for each language enable you to locate individual words at a glance. Will help language teachers and students, authors of textbooks, grammars, and language tests to compare concepts in the various languages and to concentrate on basic vocabulary, avoiding uncommon and obsolete words. 2 Appendixes. xxi + 441pp. 6½ x 9¼.
T738 Paperbound **$2.45**

NEW RUSSIAN-ENGLISH AND ENGLISH-RUSSIAN DICTIONARY, M. A. O'Brien. Over 70,000 entries in the new orthography! Many idiomatic uses and colloquialisms which form the basis of actual speech. Irregular verbs, perfective and imperfective aspects, regular and irregular sound changes, and other features. One of the few dictionaries where accent changes within the conjugation of verbs and the declension of nouns are fully indicated. "One of the best," Prof. E. J. Simmons, Cornell. First names, geographical terms, bibliography, etc. 738pp. 4½ x 6¼.
T208 Paperbound **$2.00**

96 MOST USEFUL PHRASES FOR TOURISTS AND STUDENTS in English, French, Spanish, German, Italian. A handy folder you'll want to carry with you. How to say "Excuse me," "How much is it?", "Write it down, please," etc., in four foreign languages. Copies limited, no more than 1 to a customer. **FREE**

Miscellaneous

THE COMPLETE KANO JIU-JITSU (JUDO), H. I. Hancock and K. Higashi. Most comprehensive guide to judo, referred to as outstanding work by Encyclopaedia Britannica. Complete authentic Japanese system of 160 holds and throws, including the most spectacular, fully illustrated with 487 photos. Full text explains leverage, weight centers, pressure points, special tricks, etc.; shows how to protect yourself from almost any manner of attack though your attacker may have the initial advantage of strength and surprise. This authentic Kano system should not be confused with the many American imitations. xii + 500pp. 5⅜ x 8.
T639 Paperbound **$2.00**

THE MEMOIRS OF JACQUES CASANOVA. Splendid self-revelation by history's most engaging scoundrel—utterly dishonest with women and money, yet highly intelligent and observant. Here are all the famous duels, scandals, amours, banishments, thefts, treacheries, and imprisonments all over Europe: a life lived to the fullest and recounted with gusto in one of the greatest autobiographies of all time. What is more, these Memoirs are also one of the most trustworthy and valuable documents we have on the society and culture of the extravagant 18th century. Here are Voltaire, Louis XV, Catherine the Great, cardinals, castrati, pimps, and pawnbrokers—an entire glittering civilization unfolding before you with an unparalleled sense of actuality. Translated by Arthur Machen. Edited by F. A. Blossom. Introduction by Arthur Symons. Illustrated by Rockwell Kent. Total of xlviii + 2216pp. 5⅜ x 8.
T338 Vol I Paperbound **$2.00**
T339 Vol II Paperbound **$2.00**
T340 Vol III Paperbound **$2.00**
The set **$6.00**

BARNUM'S OWN STORY, P. T. Barnum. The astonishingly frank and gratifyingly well-written autobiography of the master showman and pioneer publicity man reveals the truth about his early career, his famous hoaxes (such as the Fejee Mermaid and the Woolly Horse), his amazing commercial ventures, his fling in politics, his feuds and friendships, his failures and surprising comebacks. A vast panorama of 19th century America's mores, amusements, and vitality. 66 new illustrations in this edition. xii + 500pp. 5⅜ x 8.
T764 Paperbound **$1.65**

THE STORY OF THE TITANIC AS TOLD BY ITS SURVIVORS, ed. by Jack Winocour. Most significant accounts of most overpowering naval disaster of modern times: all 4 authors were survivors. Includes 2 full-length, unabridged books: "The Loss of the S.S. Titanic," by Laurence Beesley, "The Truth about the Titanic," by Col. Archibald Gracie; 6 pertinent chapters from "Titanic and Other Ships," autobiography of only officer to survive, Second Officer Charles Lightoller; and a short, dramatic account by the Titanic's wireless operator, Harold Bride. 26 illus. 368pp. 5⅜ x 8.
T610 Paperbound **$1.50**

THE PHYSIOLOGY OF TASTE, Jean Anthelme Brillat-Savarin. Humorous, satirical, witty, and personal classic on joys of food and drink by 18th century French politician, litterateur. Treats the science of gastronomy, erotic value of truffles, Parisian restaurants, drinking contests; gives recipes for tunny omelette, pheasant, Swiss fondue, etc. Only modern translation of original French edition. Introduction. 41 illus. 346pp. 5⅜ x 8⅜.
T591 Paperbound **$1.50**

THE ART OF THE STORY-TELLER, M. L. Shedlock. This classic in the field of effective story-telling is regarded by librarians, story-tellers, and educators as the finest and most lucid book on the subject. The author considers the nature of the story, the difficulties of communicating stories to children, the artifices used in story-telling, how to obtain and maintain the effect of the story, and, of extreme importance, the elements to seek and those to avoid in selecting material. A 99-page selection of Miss Shedlock's most effective stories and an extensive bibliography of further material by Eulalie Steinmetz enhance the book's usefulness. xxi + 320pp. 5⅜ x 8.
T635 Paperbound **$1.50**

CREATIVE POWER: THE EDUCATION OF YOUTH IN THE CREATIVE ARTS, Hughes Mearns. In first printing considered revolutionary in its dynamic, progressive approach to teaching the creative arts; now accepted as one of the most effective and valuable approaches yet formulated. Based on the belief that every child has something to contribute, it provides in a stimulating manner invaluable and inspired teaching insights, to stimulate children's latent powers of creative expression in drama, poetry, music, writing, etc. Mearns's methods were developed in his famous experimental classes in creative education at the Lincoln School of Teachers College, Columbia Univ. Named one of the 20 foremost books on education in recent times by National Education Association. New enlarged revised 2nd edition. Introduction. 272pp. 5⅜ x 8.
T490 Paperbound **$1.75**

FREE AND INEXPENSIVE EDUCATIONAL AIDS, T. J. Pepe, Superintendent of Schools, Southbury, Connecticut. An up-to-date listing of over 1500 booklets, films, charts, etc. 5% costs less than 25¢; 1% costs more; 94% is yours for the asking. Use this material privately, or in schools from elementary to college, for discussion, vocational guidance, projects. 59 categories include health, trucking, textiles, language, weather, the blood, office practice, wild life, atomic energy, other important topics. Each item described according to contents, number of pages or running time, level. All material is educationally sound, and without political or company bias. 1st publication. Second, revised edition. Index. 244pp. 5⅜ x 8.
T663 Paperbound **$1.50**

THE ROMANCE OF WORDS, E. Weekley. An entertaining collection of unusual word-histories that tracks down for the general reader the origins of more than 2000 common words and phrases in English (including British and American slang): discoveries often surprising, often humorous, that help trace vast chains of commerce in products and ideas. There are Arabic trade words, cowboy words, origins of family names, phonetic accidents, curious wanderings, folk-etymologies, etc. Index. xiii + 210pp. 5⅜ x 8.　　　　　　　　　T710 Paperbound **$1.25**

PHRASE AND WORD ORIGINS: A STUDY OF FAMILIAR EXPRESSIONS, A. H. Holt. One of the most entertaining books on the unexpected origins and colorful histories of words and phrases, based on sound scholarship, but written primarily for the layman. Over 1200 phrases and 1000 separate words are covered, with many quotations, and the results of the most modern linguistic and historical researches. "A right jolly book Mr. Holt has made," N. Y. Times. v + 254pp. 5⅜ x 8.　　　　　　　　　T758 Paperbound **$1.35**

AMATEUR WINE MAKING, S. M. Tritton. Now, with only modest equipment and no prior knowledge, you can make your own fine table wines. A practical handbook, this covers every type of grape wine, as well as fruit, flower, herb, vegetable, and cereal wines, and many kinds of mead, cider, and beer. Every question you might have is answered, and there is a valuable discussion of what can go wrong at various stages along the way. Special supplement of yeasts and American sources of supply. 13 tables. 32 illustrations. Glossary. Index. 239pp. 5½ x 8½.　　　　　　　　　T514 Clothbound **$4.00**

SAILING ALONE AROUND THE WORLD. Captain Joshua Slocum. A great modern classic in a convenient inexpensive edition. Captain Slocum's account of his single-handed voyage around the world in a 34 foot boat which he rebuilt himself. A nearly unparalleled feat of seamanship told with vigor, wit, imagination, and great descriptive power. "A nautical equivalent of Thoreau's account," Van Wyck Brooks. 67 illustrations. 308pp. 5⅜ x 8.　　　　　　T326 Paperbound **$1.00**

FARES, PLEASE! by J. A. Miller. Authoritative, comprehensive, and entertaining history of local public transit from its inception to its most recent developments: trolleys, horsecars, streetcars, buses, elevateds, subways, along with monorails, "road-railers," and a host of other extraordinary vehicles. Here are all the flamboyant personalities involved, the vehement arguments, the unusual information, and all the nostalgia. "Interesting facts brought into especially vivid life," N. Y. Times. New preface. 152 illustrations, 4 new. Bibliography. xix + 204pp. 5⅜ x 8.　　　　　　　　　T671 Paperbound **$1.50**

HOAXES, C. D. MacDougall. Shows how art, science, history, journalism can be perverted for private purposes. Hours of delightful entertainment and a work of scholarly value, this often shocking book tells of the deliberate creation of nonsense news, the Cardiff giant, Shakespeare forgeries, the Loch Ness monster, Biblical frauds, political schemes, literary hoaxers like Chatterton, Ossian, the disumbrationist school of painting, the lady in black at Valentino's tomb, and over 250 others. It will probably reveal the truth about a few things you've believed, and help you spot more readily the editorial "gander" and planted publicity release. "A stupendous collection . . . and shrewd analysis." New Yorker. New revised edition. 54 photographs. Index. 320pp. 5⅜ x 8.　　　　　T465 Paperbound **$2.00**

A HISTORY OF THE WARFARE OF SCIENCE WITH THEOLOGY IN CHRISTENDOM, A. D. White. Most thorough account ever written of the great religious-scientific battles shows gradual victory of science over ignorant, harmful beliefs. Attacks on theory of evolution; attacks on Galileo; great medieval plagues caused by belief in devil-origin of disease; attacks on Franklin's experiments with electricity; the witches of Salem; scores more that will amaze you. Author, co-founder and first president of Cornell U., writes with vast scholarly background, but in clear, readable prose. Acclaimed as classic effort in America to do away with superstition. Index. Total of 928pp. 5⅜ x 8.　　　　T608 Vol I Paperbound **$2.00**
　　　　　　　　　　　　　　　　　　　　　　　　　　　　T609 Vol II Paperbound **$2.00**

THE SHIP OF FOOLS, Sebastian Brant. First printed in 1494 in Basel, this amusing book swept Europe, was translated into almost every important language, and was a best-seller for centuries. That it is still living and vital is shown by recent developments in publishing. This is the only English translation of this work, and it recaptures in lively, modern verse all the wit and insights of the original, in satirizations of foibles and vices: greed, adultery, envy, hatred, sloth, profiteering, etc. This will long remain the definitive English edition, for Professor Zeydel has provided biography of Brant, bibliography, publishing history, influences, etc. Complete reprint of 1944 edition. Translated by Professor E. Zeydel, University of Cincinnati. All 114 original woodcut illustrations. viii + 399pp. 5½ x 8⅝.　　　　　　　　　　　　　　　　　　　　　　　　　　T266 Paperbound **$2.00**

ERASMUS, A STUDY OF HIS LIFE, IDEALS AND PLACE IN HISTORY, Preserved Smith. This is the standard English biography and evaluation of the great Netherlands humanist Desiderius Erasmus. Written by one of the foremost American historians it covers all aspects of Erasmus's life, his influence in the religious quarrels of the Reformation, his overwhelming role in the field of letters, and his importance in the emergence of the new world view of the Northern Renaissance. This is not only a work of great scholarship, it is also an extremely interesting, vital portrait of a great man. 8 illustrations. xiv + 479pp. 5⅝ x 8½.　　　　　　　　　　　　　　　　　　　　　　　　　　T331 Paperbound **$2.00**

History of Science and
Mathematics

THE STUDY OF THE HISTORY OF MATHEMATICS, THE STUDY OF THE HISTORY OF SCIENCE, G. Sarton. Two books bound as one. Each volume contains a long introduction to the methods and philosophy of each of these historical fields, covering the skills and sympathies of the historian, concepts of history of science, psychology of idea-creation, and the purpose of history of science. Prof. Sarton also provides more than 80 pages of classified bibliography. Complete and unabridged. Indexed. 10 illustrations. 188pp. 5⅜ x 8.   T240 Paperbound **$1.25**

A HISTORY OF PHYSICS, Florian Cajori, Ph.D. First written in 1899, thoroughly revised in 1929, this is still best entry into antecedents of modern theories. Precise non-mathematical discussion of ideas, theories, techniques, apparatus of each period from Greeks to 1920's, analyzing within each period basic topics of matter, mechanics, light, electricity and magnetism, sound, atomic theory, etc. Stress on modern developments, from early 19th century to present. Written with critical eye on historical development, significance. Provides most of needed historical background for student of physics. Reprint of second (1929) edition. Index. Bibliography in footnotes. 16 figures. xv + 424pp. 5⅜ x 8.   T970 Paperbound **$2.00**

A HISTORY OF ASTRONOMY FROM THALES TO KEPLER, J. L. E. Dreyer. Formerly titled A HISTORY OF PLANETARY SYSTEMS FROM THALES TO KEPLER. This is the only work in English which provides a detailed history of man's cosmological views from prehistoric times up through the Renaissance. It covers Egypt, Babylonia, early Greece, Alexandria, the Middle Ages, Copernicus, Tycho Brahe, Kepler, and many others. Epicycles and other complex theories of positional astronomy are explained in terms nearly everyone will find clear and easy to understand. "Standard reference on Greek astronomy and the Copernican revolution," SKY AND TELESCOPE. Bibliography. 21 diagrams. Index. xvii + 430pp. 5⅜ x 8.   S79 Paperbound **$2.25**

A SHORT HISTORY OF ASTRONOMY, A. Berry. A popular standard work for over 50 years, this thorough and accurate volume covers the science from primitive times to the end of the 19th century. After the Greeks and Middle Ages, individual chapters analyze Copernicus, Brahe, Galileo, Kepler, and Newton, and the mixed reception of their startling discoveries. Post-Newtonian achievements are then discussed in unusual detail: Halley, Bradley, Lagrange, Laplace, Herschel, Bessel, etc. 2 indexes. 104 illustrations, 9 portraits. xxxi + 440pp. 5⅜ x 8.   T210 Paperbound **$2.00**

PIONEERS OF SCIENCE, Sir Oliver Lodge. An authoritative, yet elementary history of science by a leading scientist and expositor. Concentrating on individuals—Copernicus, Brahe, Kepler, Galileo, Descartes, Newton, Laplace, Herschel, Lord Kelvin, and other scientists—the author presents their discoveries in historical order, adding biographical material on each man and full, specific explanations of their achievements. The full, clear discussions of the accomplishments of post-Newtonian astronomers are features seldom found in other books on the subject. Index. 120 illustrations. xv + 404pp. 5⅜ x 8.   T716 Paperbound **$1.65**

THE BIRTH AND DEVELOPMENT OF THE GEOLOGICAL SCIENCES, F. D. Adams. The most complete and thorough history of the earth sciences in print. Geological thought from earliest recorded times to the end of the 19th century—covers over 300 early thinkers and systems: fossils and hypothetical explanations of them, vulcanists vs. neptunists, figured stones and paleontology, generation of stones, and similar topics. 91 illustrations, including medieval, renaissance woodcuts, etc. 632 footnotes and bibliographic notes. Index. 511pp. 5⅜ x 8.   T5 Paperbound **$2.25**

THE STORY OF ALCHEMY AND EARLY CHEMISTRY, J. M. Stillman. "Add the blood of a red-haired man"—a recipe typical of the many quoted in this authoritative and readable history of the strange beliefs and practices of the alchemists. Concise studies of every leading figure in alchemy and early chemistry through Lavoisier, in this curious epic of superstition and true science, constructed from scores of rare and difficult Greek, Latin, German, and French texts. Foreword by S. W. Young. 246-item bibliography. Index. xiii + 566pp. 5⅜ x 8.   S628 Paperbound **$2.45**

HISTORY OF MATHEMATICS, D. E. Smith. Most comprehensive non-technical history of math in English. Discusses the lives and works of over a thousand major and minor figures, from Euclid to Descartes, Gauss, and Riemann. Vol. I: A chronological examination, from primitive concepts through Egypt, Babylonia, Greece, the Orient, Rome, the Middle Ages, the Renaissance, and up to 1900. Vol. 2: The development of ideas in specific fields and problems, up through elementary calculus. Two volumes, total of 510 illustrations, 1355pp. 5⅜ x 8. Set boxed in attractive container.   T429,430 Paperbound the set **$5.00**

A CONCISE HISTORY OF MATHEMATICS, D. Struik. A lucid, easily followed history of mathematical ideas and techniques from the Ancient Near East up to modern times. Requires no mathematics but will serve as an excellent introduction to mathematical concepts and great mathematicians through the method of historical development. 60 illustrations including Egyptian papyri, Greek mss., portraits of 31 eminent mathematicians. Bibliography. xix + 299pp. 5⅜ x 8.   T255 Paperbound **$1.75**

A SHORT ACCOUNT OF THE HISTORY OF MATHEMATICS, W. W. Rouse Ball. Last previous edition (1908) hailed by mathematicians and laymen for lucid overview of math as living science, for understandable presentation of individual contributions of great mathematicians. Treats lives, discoveries of every important school and figure from Egypt, Phoenicia to late nineteenth century. Greek schools of Ionia, Cyzicus, Alexandria, Byzantium, Pythagoras; primitive arithmetic; Middle Ages and Renaissance, including European and Asiatic contributions; modern math of Descartes, Pascal, Wallis, Huygens, Newton, Euler, Lambert, Laplace, scores more. More emphasis on historical development, exposition of ideas than other books on subject. Non-technical, readable text can be followed with no more preparation than high-school algebra. Index. 544pp. 5⅜ x 8.   S630 Paperbound **$2.00**

ON MATHEMATICS AND MATHEMATICIANS, R. E. Moritz. A ten year labor of love by the discerning and discriminating Prof. Moritz, this collection has rarely been equalled in its ability to convey the full sense of mathematics and the personalities of great mathematicians. A collection of anecdotes, aphorisms, reminiscences, philosophies, definitions, speculations, biographical insights, etc., by great mathematicians and writers: Descartes, Mill, De Morgan, Locke, Berkeley, Kant, Coleridge, Whitehead, Sylvester, Klein, and many others. Also, glimpses into the lives of mathematical giants from Archimedes to Euler, Gauss, and Weierstrass. To mathematicians, a superb book for browsing; to writers and teachers, an unequalled source of quotation; to the layman, an exciting revelation of the fullness of mathematics. Extensive cross index. 410pp. 5⅜ x 8.   T489 Paperbound **$1.95**

SIR ISAAC NEWTON: A BIOGRAPHY, Louis Trenchard More. Standard, definitive biography of Newton, covering every phase of his life and career in its presentation of the renowned scientific genius as a living man. Objective, critical analysis of his character as well as a careful survey of his manifold accomplishments in many areas of science, and in theology, history, politics, finance. Text includes letters by Newton and acquaintances, many other papers, some translated from Latin to English by the author. Scientists, teachers of science will especially be interested in this book, which will appeal to all readers concerned with history of ideas, development of science. Republication of original (1934) edition. 1 full-page plate. Index. xii + 675pp. 5⅜ x 8½.   S79 Paperbound **$2.50**

GUIDE TO THE LITERATURE OF MATHEMATICS AND PHYSICS, N. G. Parke III. Over 5000 entries included under approximately 120 major subject headings, of selected most important books, monographs, periodicals, articles in English, plus important works in German, French, Italian, Spanish, Russian (many recently available works). Covers every branch of physics, math, related engineering. Includes author, title, edition, publisher, place, date, number of volumes, number of pages. A 40-page introduction on the basic problems of research and study provides useful information on the organization and use of libraries, the psychology of learning, etc. This reference work will save you hours of time. 2nd revised edition. Indices of authors, subjects. 464pp. 5⅜ x 8.   S447 Paperbound **$2.49**

Dover publishes books on art, music, philosophy, literature, languages, history, social sciences, psychology, handcrafts, orientalia, puzzles and entertainments, chess, pets and gardens, books explaining science, intermediate and higher mathematics, mathematical physics, engineering, biological sciences, earth sciences, classics of science, etc. Write to:

> *Dept. catrr.*
> *Dover Publications, Inc.*
> *180 Varick Street, N. Y. 14, N. Y.*